TRANSACTIONS

OF THE

AMERICAN PHILOSOPHICAL SOCIETY

HELD AT PHILADELPHIA
FOR PROMOTING USEFUL KNOWLEDGE

NEW SERIES—VOLUME 52, PART 3
1962

STUDIES ON THE MARINE PLEISTOCENE:

PART I. THE MARINE PLEISTOCENE OF THE
AMERICAS AND EUROPE

PART II. THE MARINE PLEISTOCENE MOLLUSKS
OF EASTERN NORTH AMERICA

HORACE G. RICHARDS

Academy of Natural Sciences of Philadelphia

THE AMERICAN PHILOSOPHICAL SOCIETY
INDEPENDENCE SQUARE
PHILADELPHIA 6

JULY, 1962

Library of Congress Catalogue
Card No. 62-18221

PREFACE

Studies on the marine Pleistocene of the Atlantic Coast of North America were carried out at various times between approximately 1931 and 1938 and intermittently thereafter. The field work between 1934 and 1938 was aided by grants from the Geological Society of America, and resulted in several publications (MacClintock and Richards, 1936; Richards, 1936, 1938, 1939a,b). While lists of species were given, no descriptions, synonomies, or illustrations were included in these reports.

Studies on the marine Pleistocene of James Bay were made in 1933, on the west coast of Hudson Bay in 1940, and in the Mackenzie Delta in 1948, the latter under a grant from the American Philosophical Society, while shorter collecting trips have been made to the post-glacial marine deposits of New England, the St. Lawrence Valley, and the Lake Champlain region.

Brief investigations of the Pleistocene of certain of the islands of the Caribbean were made at various times between 1933 and 1960. Three trips (to Cozumel, 1936; Roatan, 1937; and Corn Island, 1938) were aided by grants from the American Philosophical Society, while one trip (mainly to Margarita Island, off Venezuela) was under a grant from the Geological Society of America.

Studies, frequently purely of a reconnaissance nature, have been carried out on other islands of the West Indies, including Cuba (1933), Grand Cayman (1952), Bahamas (1937, 1959), Virgin Islands (1960), Puerto Rico (1959), St. Martins and Anguilla (1959).

A reconnaissance trip to Peru and Ecuador was made in June, 1961, with the cooperation of the International Petroleum Company, Ltd. (Talara, Peru), and the Anglo-Ecuadorian Oilfields (Guayaquil, Ecuador).

While no pretense has been made to undertake detailed studies of the marine Pleistocene in the other places mentioned in this report, the writer has had the opportunity to make brief visits of most of the areas discussed including the following: California and Baja California, 1941, 1946, 1961; East Anglia (England), 1949; Italy (near Rome), 1953, 1954; North Africa (Algeria, Tunisia), 1953; Lebanon, 1954; Israel, 1955; Spain, 1957; Scandinavia, 1956, 1960; Iceland, 1960.

The present report is in two parts. The first six chapters summarize our knowledge of the marine Pleistocene shore lines and fossil deposits of North and South America and Europe, while the second part records and figures the marine Pleistocene mollusks of the Atlantic Coast of North America between Hudson Bay and Georgia.

It was originally planned to prepare an illustrated report on the marine Pleistocene mollusks of the entire East Coast of North America from Hudson Bay to Florida. However, it was decided to limit the study to the coast of Georgia northward. The Florida fauna, Recent and Pleistocene, is very rich, and at least the southern part is more closely related to the fauna of the West Indies than to that of the rest of the Atlantic Coast. To have included the Florida species would have enlarged the study considerably and was not practical at the present time.

The preparation of some of the illustrations as well as some supplemental field work was made possible by a grant from the Johnson Fund of the American Philosophical Society.

A grant from the Faculty Research Fund of the University of Pennsylvania in 1961 was used for some final field work as well as for some additional photographs, retouching, and the preparation of the sketch maps.

The entire manuscript of Part I has been critically read by Paul MacClintock of Princeton University and William Farrand of Columbia University, and it is a pleasure to thank these gentlemen for their many helpful suggestions.

Individual sections of the manuscript have been read by various specialists or students in the various areas discussed, and I am indebted to the following for their comments: Edwin C. Allison (San Diego State College, Calif.), J. J. Donner (University of Helsinki, Finland), Rhodes Fairbridge (Columbia University, New York), Albert Fischer (Princeton University), Bryce M. Hand (then at University of Southern California, Los Angeles, now at Pennsylvania State University), R. L. Merklin (Paleontological Institute, Moscow, U.S.S.R.), Walter Newman (Queens College, Flushing, N. Y.), Axel Olsson (Miami, Fla.), W. Armstrong Price (Corpus Christi, Texas), Edward Sammel (U. S. Geological Survey, Boston), Andrei Sinizin (University of Leningrad, U.S.S.R.), A. van der Heide (Geological Survey of the Netherlands, Haarlem), A. L. Washburn, (Yale University, New Haven).

R. Tucker Abbott of the Academy of Natural Sciences of Philadelphia critically read the paleontological portions of the paper (Part II) and offered many helpful suggestions. Others who have advised on particular genera include J. Lockwood Chamberlin (U. S. Fish and Wildlife Service, Washington, D. C.) and F. Stearns MacNeil (U. S. Geological Survey, Menlo Park, Calif.).

Wallace Broecker of the Lamont Geological Observatory, Palisades, N. Y., supplied radiocarbon dates for some of the shells.

The bibliography appears at the end of Part II.

H. G. R.

3

NOTE

Since this report deals with many different parts of the world, the two main units of measurement are encountered—metric (meters and kilometers) and English (feet and miles). Instead of translating to a single system, it has been decided to retain the units used by the various authors cited in the text. It is thought that this will create less confusion than changing feet and miles to meters and kilometers in an area where the English system is in general use.

To translate from one system to the other, the formula 1 meter = 3.3 feet may be used.

STUDIES ON THE MARINE PLEISTOCENE

Part I. The Marine Pleistocene of the Americas and Europe
Part II. The Marine Pleistocene Mollusks of Eastern North America

HORACE G. RICHARDS

CONTENTS

PART I

THE MARINE PLEISTOCENE OF THE AMERICAS AND EUROPE

I. THE PLEISTOCENE EPOCH

The complexity of the Pleistocene epoch (or Quaternary period) has long been known, and numerous attempts have been made to subdivide it and to correlate the subdivisions of different parts of the world. It is generally believed that the main glacial and interglacial stages of North America and Europe were contemporaneous, although there is some doubt about the older stages.

Table 1 shows the main subdivisions of the Pleistocene as generally recognized in North America, the Alps, and Northern Germany. The terminology of

North America and the Alps is that used by most recent workers (see Flint, 1957; Zeuner, 1959, etc.). That from Northern Germany is largely from Woldstedt (1955). As indicated by the "?" in the table, the correlations of the earlier stages are not firmly established.

In New Jersey and Pennsylvania an old (pre-Wisconsin) drift was named the Jerseyan. Its exact correlation is uncertain, but part of it may be Illinoian and part Kansan (see MacClintock and Richards, 1936: 304–305; MacClintock, 1940: 113).

Correlations with other parts of the world are less

TABLE 1

CORRELATION TABLE OF THE PLEISTOCENE OF NORTH AMERICA, THE ALPS, AND NORTHERN GERMANY

U.S.A.	Alps	North Germany
Wisconsin glacial stage	Würm glacial	Weichsel glacial
Sangamon interglacial	Riss/Würm interglacial	Eem interglacial
Illinoian glacial	Riss glacial	Saale glacial
Yarmouth interglacial	Mindel/Riss interglacial	Holstein marine beds "Great interglacial"
Kansan glacial	Mindel glacial	Elster glacial
Aftonian interglacial	Günz/Mindel ? interglacial	Cromer interglacial
Nebraskan ? interglacial	Günz glacial	? Weybourne "kaltzeit"
	Donau/Günz "warmzeit" ?	Tegelen "warmzeit"
	Donau glacial?	? Brachtier "kaltzeit"

definite, but some will be discussed later in this volume. Table 1 may serve as a useful reference for later discussions.

It has been recognized that the various glacial and interglacial stages, shown in table 1, can be divided into subunits indicating fluctuations in climate. The most detailed work has been done on the last stage (Wisconsin = Würm). Table 2 shows the subdivisions of the Main (or Classical) Wisconsin in north-central United States according to Leighton (1960).

An alternate proposal for the subdivision of the Wisconsin has been made by Frye and Willman (1960) for the region of the Lake Michigan glacial lobe. This is reproduced in figure 1 (from Frye and Willman, 1960: 2, 3). "The Wisconsinan is plotted at 55,000 years B.P., however it is judged to be at least 50,000 and perhaps as much as 70,000 B.P." [1] It will be noted that half of Wisconsin(an) time falls within the Altonian, the oldest of the substages. Under former classifications, the Sangamon was considered to end at about 28,000 years B.P.

TABLE 2

SUBSTAGES OF THE MAIN WISCONSIN (AFTER LEIGHTON)

MAIN WISCONSIN

Valders (Glacial)
Two Creeks (intraglacial)
Mankato (glacial) [2]
Bowmanville (intraglacial)
Cary (glacial)
Tazewell (glacial)
Gardena (intraglacial)
Iowan (glacial)
Farm Creek (intraglacial)
Farmdale (glacial)

[1] Before the present.
[2] The Mankato is called Port Huron east of Lake Michigan.

It will be noted that Leighton (1960) uses the terms "glacial" and "interglacial" for the substages of the Wisconsin stage, while Frye and Willman (1960) merely use the term "substage" for each unit. Other writers prefer the terms "stadial" and "interstadial" for the respective substages. Since the two systems of subdividing the Wisconsin are incompatible, no attempt will be made here to pass judgment on either.

Some writers use the term Brady for an interstadial soil which apparently lies between the Tazewell and Cary (see Horberg, 1955).

Attempts have been made to correlate the Wisconsin subdivisions of North America with those of Europe, but the only correlation widely accepted is that of the Two Creeks interstadial with the Alleröd of Europe, both of which have dated by radiocarbon between 11,000 and 12,000 years B.P. In fact, the various substages of the Wisconsin of the north-central stages shown in table 2 have not yet been correlated definitely with substages in the northeastern states.

Recently it has been shown that there are glacial and interglacial deposits in the Great Lakes area that antedate the Main Wisconsin, but which postdate the Sangamon interglacial. These have been termed Early and Mid-Wisconsin in contrast to the Main or "Classical" Wisconsin (Dreimanis, 1960), or simply "Early Wisconsin" (Goldthwait, 1958).

According to Dreimanis (1960), the Early Wisconsin ice advanced, with oscillations, more than 66,000 years B.P. over southern Quebec and southwestward. Its retreat was followed by a long, cool, Mid-Wisconsin (Sidney) interstadial. About 44,000 years ago a Mid-Wisconsin glacial advance reached at least as far as Lake Erie, dividing the interstadial into two parts (Port Talbot and Plum Point). This was followed about 30,000 years ago by the beginning of the Main or "Classical" Wisconsin, or the "Late Wisconsin" of Goldthwait (1958). The units of the Early and Mid-Wisconsin are shown in figure 2 (from Dreimanis).

Radiocarbon dates between about 44,000 and 47,000 years B.P. for the Port Talbot interstadial, and between about 24,600 and 28,200 years B.P. for the Plum Point interstadial, have been obtained (deVries and Dreimanis, 1960).

The terms "late-glacial" and "postglacial" are widely used, but frequently are given different meanings by different authors. The term "late-glacial" is used especially in Europe and is sometimes taken to indicate the time between the disappearance of the ice sheets and the final appearance of the forests. However, it is obvious that this interval would have occurred at different times in different places. Therefore, the term "late-glacial'" has little chronologic significance and will generally be avoided in this report except in quotations.

The term "postglacial" also has different chronologic meaning depending upon (1) location and (2) the particular phase of the Wisconsin referred to. While there

ILLINOIS STATE GEOLOGICAL SURVEY   CLASSIFICATION OF THE WISCONSINAN

RECENT STAGE	(Alluvium)	SELECTED RADIOCARBON DATES FROM ILLINOIS (B. P.)			RECENT STAGE

RECENT STAGE — (Alluvium) — 5,000

VALDERAN SUBSTAGE — (Alluvium)

6,340 ± 250 (W-317) alluvium, Miss. R. Valley

— 11,000 — VALDERS
TWOCREEKAN SUBSTAGE
— 12,500 —

10,700 ± 300 (W-426) Lake Chicago
[Wisconsin - 11,400 (Av. of dates)]
Two Creeks Forest bed

WOODFORDIAN SUBSTAGE — Peoria loess
(Alluvium and lacustrine deposits) — LAKE BORDER MORAINES
VALPARAISO MORAINES
Richland loess — MARSEILLES MORAINES
CROPSEY MORAINES
BLOOMINGTON - NORMAL MORAINES
LEROY - CHAMPAIGN MORAINES
SHELBYVILLE MORAINES
Morton loess

15,600 ± 600 (W-381) terrace silts, Ill. R. Valley

17,100 ± 300 (W-730) Peoria loess
[Missouri - 17,000 ± 600 (W-470) terrace]

19,200 ± 700 (W-187) Shelbyville till
20,340 ± 750 (W-349) Morton till
20,700 ± 650 (W-399) Morton loess

— 22,000 — ?   ? ?

FARMDALIAN SUBSTAGE — Farmdale silt and peat

22,900 ± 900 (W-68) Farmdale silt
23,500 ± 400 (W-745) peat, Ill. R. terrace
23,550 ± 550 (W-849) Farmdale peat
25,100 ± 800 (W-69) Farmdale silt
25,500 ± 600 (W-853) Farmdale peat
26,100 ± 600 (W-381) Farmdale peat

— 28,000 —
Winnebago drift

[Wisconsin - 31,800 ± 1200 (W-638) till]
31,000 or older (W-186) till, Bloomington, Ill.

(Inferred southern limit of glacial ice in the Lake Michigan lobe.)

35,200 ± 1,000 (W-729) Roxana loess

37,000 ± 1,500 (W-869) Roxana loess

ALTONIAN SUBSTAGE — Roxana silt — (Alluvium)

[Iowa - >35,000 (W-516) Iowan till]

>37,000 (W-256) till, Danville, Ill.
38,000 (W-243) alluvium, Miss. R. Valley

— 50,000 - 70,000 —
SOIL

WISCONSINAN STAGE

FORMER CLASSIFICATION:
RECENT STAGE

WISCONSIN STAGE:
MANKATO
CARY
TAZEWELL
IOWAN
FARMDALE

SANGAMON STAGE

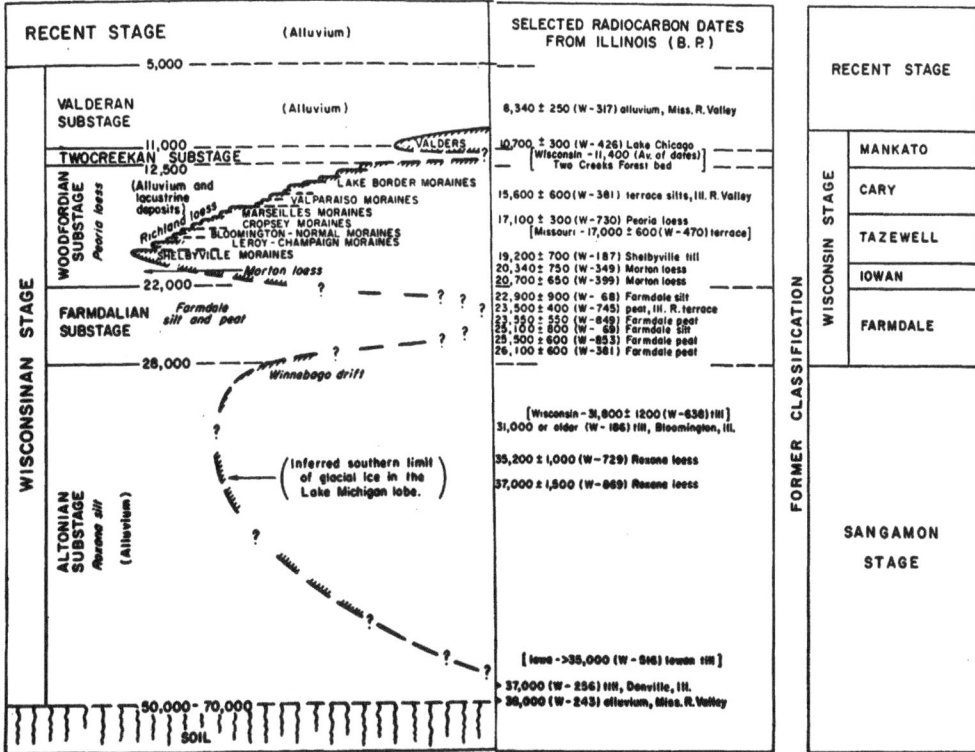

FIG. 1. Time-stratigraphic subdivisions of Wisconsinan Stage in the Lake Michigan glacial lobe (after Frye and Willman, 1960).

is not full agreement, many geologists place the beginning of the postglacial—or Holocene as it is frequently called by European geologists—at about 11,000 years B.P. (Broecker *et al.*, 1960; Fairbridge and Newman, 1961).

In this report the term "postglacial" will be used only in a general sense, as was done by Flint (1957), mainly to differentiate such deposits from those of interglacial age.

II. MARINE PLEISTOCENE

Studies on the marine Pleistocene of Eastern North America illustrate three well-established principles. One of these is that of glacial control of sea level; it has been estimated that, at the climax of the last (Wisconsin-Würm) glaciation, enough water was removed from the sea to have appreciably lowered sea level at least 300 feet (100 meters) or possibly as much as 450 feet (150 meters). Conversely, during the mild interglacial stages, sea level was higher than now, and it

has been estimated that, during the last interglacial (Sangamon-Riss/Würm), sea level was between 25 and 50 feet (8 and 17 meters) higher than at present.

The second hypothesis concerns the effect of the weight of the glacial ice on the land. It is generally believed that portions of North America, as well as

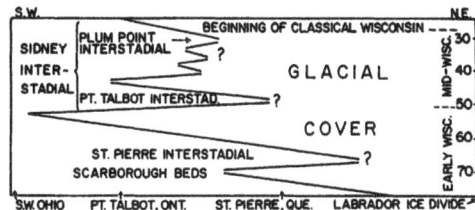

FIG. 2. Tentative stratigraphy of the pre-classical Wisconsin of the Great Lakes region. Figures at right indicate thousands of years before the present (after Dreimanis, 1960).

other glaciated areas, were considerably depressed by the weight of the ice, more to the north where the quantity of ice was greater. As the glaciers began to melt, water was poured back to the sea, causing the sea to flood the land which had been pushed down by the ice. Gradually the land recovered from the weight of the ice, and slowly rose. The total amount of crustal uplift recorded is equivalent to the highest elevation at which marine features can be found plus an amount equal to the rise of sea level since the marine limit was formed. Thus, the marine limit indicates the excess of the rise of the land over that of the sea since the retreat of the ice sheets.

There was also an unknown but large amount of uplift before deglaciation, that is before the marine limit was formed.

A third factor to be considered is the diastrophic movement of the land independent of any effect of the ice. Such movements played an important part in the Pleistocene history of many parts of the world, for example, along the west coast of North and South America and certain of the islands of the West Indies. On the other hand, diastrophic movements apparently played a relatively insignificant role in the Pleistocene history of the east coast of North America.

In summary, therefore, north of the terminal moraine (near New York City) one finds evidence that the coastal land was invaded by a cold, shallow sea shortly after the retreat of the Wisconsin ice. South of the moraine (Long Island to Florida) the most recent transgression of the sea was interglacial (Sangamon). Records of the interglacial sea have largely been obliterated north of the terminal moraine, although traces are known from Nantucket Island (Massachusetts) and possibly elsewhere.

The cold sea which covered portions of northeastern North America as the ice was melting is sometimes called the "Champlain Sea." This sea extended up the St. Lawrence Gulf and River to Montreal with arms projecting north along the Ottawa River beyond Ottawa, east along the St. Lawrence almost to the Thousand Islands, and south into Lake Champlain. The sea also covered the land adjoining the shores of Hudson and James Bays (Canada) where it is called the "Tyrrell Sea" (Lee, 1960), parts of Newfoundland and Labrador as well as the coastal regions of Maine and New Hampshire.

The post-Wisconsin molluscan fauna consists of about fifty-five species of pelecypods, and fifty-one gastropods (omitting certain minute genera). Many of these species occur also in late- and postglacial deposits of Norway and Sweden and the Kola Peninsula of the U.S.S.R., although at the latter locality some of them are probably of interglacial age.

In Scandinavia the detailed history of the warping of the land as well as the rise of sea level together with the succession of the faunas has long been known. A similar history took place in the Champlain Sea of North America.

South of the terminal moraine, in other words from Long Island to Florida, we find excellent evidence of a warm interglacial sea 28 feet (9 meters) above present sea level. This is the Cape May or Pamlico formation. There is also physiographic evidence of a shore line at 90 feet (30 meters) called the Surry Scarp. The lower shore line and questionably the higher one are dated from the Sangamon interglacial.

Terrace deposits above the 30-meter level have been recognized by Cooke (1930, 1931, 1932, 1935) and others along the South Atlantic Coastal Plain and have been regarded as of marine origin, dating from the earlier interglacial stages of the Pleistocene. Other geologists, notably Flint (1940b, 1942), Hack (1955), and Richards (1936), have questioned the marine origin of some of these higher terraces and have preferred to regard them as of fluvial origin. One possible exception is the presence of early Pleistocene (?) fossils at elevation 65 feet in South Carolina. However, as will be pointed out later (p. 17), the fauna may be of late Pliocene age. Shore-line features have also been reported from Georgia and Florida above the Surry Scarp, that is, above 90 feet (MacNeil, 1950). However, in view of the lack of marine fossils and positively identified shore lines, the existence of high-level, older, Pleistocene shore lines along the east coast of North America must be regarded as still unproved.

The interglacial fauna indicates a climate at least as warm as that of today in the same latitudes, and frequently somewhat warmer. The fauna suggests shallow water (probably less than 60 feet = 20 meters) and is characterized by the absence or extreme scarcity of deep-water species.

The interglacial fauna from Long Island to Georgia consists of about 140 pelecypods and 90 gastropods, again omitting certain smaller genera, that have little stratigraphic usefulness. As might be expected, very few of the species of this interglacial fauna are common to the deposits of North America and Europe.

III. SUMMARY OF PLEISTOCENE HISTORY BY AREAS

In this section an attempt will be made to give brief summaries of the Pleistocene history of the shore lines of North, Central, and South America, the Caribbean Islands, and Europe with brief notes on the Siberian coast and that of North Africa. No attempt will be made to summarize the literature completely or to settle controversial opinions. However, it is hoped that it will be possible to stress the various factors that have influenced the Pleistocene history of the various shore lines.

HUDSON BAY AND EASTERN CANADIAN ARCTIC

Elevated strand lines, beach ridges, and other marine features are widely distributed on the mainland of northern Canada, especially along the shore of Hudson Bay, as well as in the Arctic Archipelago. These features indicate considerable depression of the land caused by the weight of the Wisconsin ice, and subsequent uplift. The faunas in all cases indicate cold water, and are thought to date from post-Wisconsin time. Studies on these faunas and marine features have been made by O'Neill (1924), Nichols (1936a,b), Richards (1941), and many others.

A few typical localities, showing elevated beaches or other marine features, together with elevation, are listed below. Fossils are reported from localities marked *. The features of those localities marked ? are questionably marine. These localities do not necessarily represent the marine limit and are frequently considerably below it.

Locality	Elevation (in feet)	Reference
"Baffin Island"	1300?	O'Neill (1924); Nichols (1936a)
"Southampton Island"	1300?	O'Neill (1924); Nichols (1936a)
*Craig Harbor, Ellesmere I.	10	Nichols (1936a)
*Wolstenholme	550	Nichols (1936a)
Hudson Strait	898	Nichols (personal communication)
*Sugluk, Que.	224	Nichols (1936a)
*Port Harrison, Que.	10–162	Nichols (1936a)
*Baker Lake, N.W.T.	100	Richards (1941)
*Chesterfield Inlet, N.W.T.	(a few feet)	Richards (1941)
*Churchill, Manitoba	80	Nichols (1936a); Richards (1941)
*Victoria Island, N.W.T.	510 (526?)	Washburn (1947)
*Belcher Islands, N.W.T.	300	Richards (1940a)
Frobisher Bay, N.W.T.	1425?	Mercer (1956)
Cornwallis and Devon Island	1000?	Sutherland (1853)
N. Ellesmere Island	2000?	Greeley (1888)
Richmond Gulf	875	Stanley (1939)

Much additional information is included in the works of Lee (1959), Bird (1959), Dunbar (1959), Craig and Fyles (1961), and Farrand and Gajda (1961,1962).

It is highly probable that uplift is continuing today since there is frequent evidence of ancient Eskimo ruins, probably originally built directly on the shore, and now as much as 80 feet above tide level and as much as 3,900 feet from the present shore line (Nichols, 1936b: 6).

The Baker Lake localities are of particular interest. The presence of elevated beaches with shells 100 feet above the level of the lake, and 225 miles from Hudson Bay suggests a rather extensive submergence of the "Barren Lands" in this area. Similar elevated beaches have been observed along Wager Bay, some 150 miles north of Chesterfield Inlet. It seems highly probable that there was a direct marine connection between one

or more of these inlets of Hudson Bay and an arm of the Arctic Ocean, now separated by more than 100 miles of low "Barren Lands."

A map showing the isobases of maximum postglacial submergence has recently been prepared by Farrand and Gajda (1961, 1962). This shows a maximum submergence along the east coast of Hudson Bay of 900 feet.[3]

During times of glaciation when sea level was low, it is most probable that the Arctic Ocean was more or less completely isolated from the Atlantic as well as from the Pacific (Dunbar, 1959: 56–57).

JAMES BAY, CANADA

James Bay has had a history very similar to that described for Hudson Bay (Kindle, 1924; Richards, 1936). Pleistocene marine deposits have been found along the Moose River and tributaries as well as other rivers that flow into James Bay. These are thought to be of post-Wisconsin age deposited as the Wisconsin glaciers withdrew to the north and when the land was low because of the weight of the ice. The release of

FIG. 3. Shell ridge at western end of Baker Lake, N.W.T.

the load of the ice caused the land to rise and the sea to withdraw to the north. A layer of silt, exposed along the Moose River near Moose Factory containing fresh-water and land mollusks, is evidence for this stage of withdrawal (Richards, 1936: 541).

That James Bay was deeper and more saline than at present is shown on the beaches of Charlton and Cary islands (85 miles north of Moose Factory) by the finding of numerous shells that are not living in James Bay today. It is thought that these shells are of late Wisconsin age (Richards, 1936b: 534).

Recent studies of Hudson and James Bays by Lee (1960) have demonstrated the extent of the postglacial submergence in this area. This sea is thought to have reached its maximum extent about 7,000 to 8,000 years ago (based upon radiocarbon studies) and has been named the "Tyrrell Sea" in honor of the explorer-geologist Joseph Burr Tyrrell who was one of the first to recognize this submergence.

---

[3] For discussion of the Western Canadian Arctic see pages 24, 25.

FIG. 4. Silt containing land and fresh-water mollusks overlain by nonfossiliferous muskeg. The silt overlies a clay containing a brackish fauna (not exposed). Moose River, Ont.

Lougee (1958) believed that massive brown clays in the vicinity of Cochrane, Ontario, were of marine origin and represented a southern extension of James Bay. These clays occur at elevation 1,000 feet and 200 miles south of present James Bay. However, the absence of fossils casts some doubt on the marine origin of the clay.

A possible late Pleistocene sea connection between James Bay and the Gulf of St. Lawrence via the Ottawa River and the Lake Temiskaming region was suggested by Potter (1932) on the basis of the distribution of certain living halophytes. The exact route of the sea connection was objected to on geologic grounds by LaRocque (1949) who proposed a possible alternate route via the Saguenay River, Lake St. John, and adjacent lowlands. It must, however, be emphasized that the botanical evidence has not been confirmed by the finding of marine fossils between Lake St. John and the lower tributaries of James Bay. However, as pointed out by LaRocque, the region has not been thoroughly explored.

LABRADOR AND NEWFOUNDLAND

Marine fossils have been reported from elevated beaches along the coast of Labrador up to at least 265

feet above sea level, and marine features have been noted up to 390 feet (Packard, 1865; Daly, 1902).[4]

Apparently at least the north and west coast of Newfoundland was covered by the late Wisconsin sea. Emerged features occur along the west coast increasing in elevation to the northwest to about 250 feet above sea level (Flint, 1940b). Apparently eastern Newfoundland is tilting up to the northwest. The zero isobase—or hinge line—passes through Trinity and Placentia Bays according to Jenness (1960), although Farrand and Gajda (1961) place it in the vicinity of St. Johns. Northwest of this axis, the coast is emerging, while to the southeast it is submerging (Jenness, 1960).

Thirty-one species of mollusks were listed from some forty-one fossil localities, all indicating a cold, shallow sea, and dated from the late Wisconsin submergence (Richards, 1940b).

Coleman (1926) favored the concept of multiple glaciation, and believed that at least some of the fossil beds were of interglacial age. However, this interpretation has been refuted by more recent investigators (MacClintock and Twenhofel, 1940).

Emerged marine features have been reported from the Maritime Provinces (Nova Scotia and New Brunswick by Goldthwait (1924) and others) and fossils have been reported at a few places.

ST. LAWRENCE GULF AND RIVER

This same late- or postglacial sea extended up the valley of the St. Lawrence one hundred miles west of Montreal. Considerable depression and subsequent elevation are indicated because marine fossils have been found at the following elevations:

Rivière du Loup, Que.	300 feet
Quebec, Que.	180 feet
Montreal, Que.	560 feet
Ottawa, Ont.	510 feet

The most extensive fauna is that from Rivière du Loup (Dawson, 1871) and contains many species suggestive of relatively deep, cold water. Coleman (1927, 1941) believed that there was evidence of two glaciations in the St. Lawrence region with an interglacial, marine shell bed between. However, this interpretation is not held by more recent workers, and the prevailing opinion among geologists today is that all the shells from the Pleistocene beaches of the St. Lawrence area date from late Wisconsin and post-Wisconsin time. See B. Smith (1947) and others.

Arms of this sea extended north along the Saguenay River lowland to Lake St. John (Tolmachoff, 1927), north beyond Ottawa at least as far as Lake Coulonge,

[4] Bell (1890: 308) reported raised beaches at Nachvak in Northern Labrador 1,500 feet above the sea. However, this elevation is in disagreement with data of Daly (1902).

seventy miles west of that city (Johnston, 1916), south into the Lake Champlain lowland, as well as west to the vicinity of the Thousand Islands. The fauna of these various arms of the Champlain Sea definitely suggests shallow brackish water. At Prescott, Ontario (near the Thousand Islands), there is a mixture of species of fresh- and brackish-water affinity. One locality near Ottawa (Green Creek) is noted for the presence of fossil fish (*Mallotis villusus*, *Cyclopterus* sp., and *Gasterosteus* sp.) in concretions.

Antevs (1939) and other workers have suggested the possibility of two late Pleistocene marine invasions of the Champlain Sea, especially in the Ottawa area. The latter invasion was called the "Ottawa Sea." However, more recent work has suggested that the land was submerged only once and has been gradually uplifted (see summary by Gadd, 1961).

LAKE CHAMPLAIN LOWLANDS, NEW YORK, AND VERMONT

Some fifteen fossil localities have been reported along the west (New York) side of Lake Champlain and twelve on the east (Vermont) side (Goldring, 1922; Howell and Richards, 1937). A dwarfing of the fauna was noted in the southern part of the embayment (Shimer, 1908). On the other hand, fossil whale remains have been found in the Champlain Sea at Smith Falls, Ontario.

While the Champlain Sea of the St. Lawrence Lowlands has in the past generally been regarded as late post-Wisconsin, recent geological investigations coupled with palynological and radiocarbon studies suggest that part of the Champlain Sea episode may date from the Two Creeks Interstadial (Terasmae, 1959b).

MacClintock and Terasmae (1960) have shown that sea level began to rise about 12,000 years B.P. and that the marine deposits of the Champlain Sea overlie the Fort Covington till (Port Huron age). They cite radiocarbon dates from Champlain Sea shells between 10,300 and 11,300 years B.P. The basal part of terrestrial peat on top of Champlain Sea sediments near Drommondville, Quebec, has been dated at 9,500 years B.P. (Terasmae, 1959b). Thus the Champlain Sea of the St. Lawrence Lowlands is thought to have existed about 11,300 B.P. to about 9,500 years B.P., and is thus correlated with part of Two Creeks time (see tables on pages 6, 7).

The contemporaneity of Early Man with the presence of the Champlain Sea is pointed out by Mason (1960). His analysis of the data would allow Paleo-Indian entry into northeastern North America by 11,000 to 10,500 years ago.

MAINE

Postglacial marine shells have been found along the coast of Maine up to elevation 300 feet. In some places the shells occur in clay lying on top of eskers, thus demonstrating their post-Wisconsin age. Shell beds have been reported by Packard (1865), Clapp (1907), Leavitt and Perkins (1935), and others. The species are largely of northern affinity and frequently also occur in the late Pleistocene deposits of Scandanavia. In addition to mollusks, fossil birds, fish, seals, walrus, and whales have been reported (Leavitt and Perkins, 1935: 205–207).

Clapp favored the concept of multiple glaciation in Maine, and believed that some of the shell deposits antedated the Wisconsin. However, current opinion is that the apparent differences in weathering can best be explained by differences in the bedrock, and therefore all glaciation should be assigned to the Wisconsin (Leavitt and Perkins, 1935: 183–184).

Bloom (1960) has recently made extensive studies on the Late Pleistocene deposits of southwestern Maine, and has listed marine fossils from some thirty-three localities. He has given the name "Presumpscot formation" to the sheet of emerged marine sediments that blankets the coastal region of southwestern Maine. Although marine submergence has been demonstrated by fossil evidence to an altitude of more than 260 feet, probably to 280 or 290 feet, the shore-line features, such as wave-cut cliffs, bars, and deltas, are distributed through considerable range in present altitude. This may be explained by rapid submergence and reemergence, and is in line with the hypothesis of an abrupt change in climate about 11,000 years ago (Broecker, Ewing, and Heezen, 1960).

According to Bloom, the maximum submergence of southwestern Maine probably took place about 11,800 years ago as indicated by radiocarbon dates of marine shells from Waterville, Maine. The subsequent emergence caused the land to extend beyond the present shore line, as indicated by pollen studies and by the presence of numerous buried valleys and trees close to the present shore. A drowned forest at Robin Hood,

FIG. 5. Fossiliferous Presumpscot formation overlain by till, near Portland, Maine.

Maine, had been dated at $4,150 \pm 200$ years B.P. Transition to present conditions was caused mainly by the eustatic rise of sea level, although there is some evidence for a recent subsidence of the land.

Bloom proposed that an ice lobe (Kennebunk) pushed, as an active advance, into the sea. Such an advance would be the same age as the St. Narcisse moraine of Quebec, formed when the Valders ice pushed into the northern edge of the Champlain Sea.

MASSACHUSETTS

Stimpson (1851) lists fourteen species of shells from the Quaternary of Point Shirley, Mass., about 50 feet above high water. All the species are known to be living in Massachusetts waters today.

According to Horner (1929: 411):

When the region was released from the ice, the shore line in northeastern Massachusetts lay higher than at present. It withdrew rapidly to below the present strand, reaching in the Boston region at least the level of minus 43 feet. Finally it transgressed to its present position.

Horner found no trace of a postglacial invasion of the sea south of Boston, and therefore we may conclude that marine fossils found south of that point belong to an older stage (interglacial).

Shimer (1918) recorded dark-gray silt with a rich molluscan fauna resembling that now living off the coast of Virginia, some 13 to 15 feet below the surface level of Boyleston and Berkeley Street in Boston. These were associated with an ancient fish wier. He estimated the age to be about 2,000 to 3,000 years. Others have suggested that these beds date from the postglacial temperature maximum (Hypsithermal time) which occurred from about 7,500 to 4,000 years ago. On the other hand, after reviewing the geologic evidence, Judson (1949) favors the belief that the fish weir postdates Hypsithermal time and may date from anywhere between 2,000 and 3,000 years before the present.

At Sankaty Head on the island of Nantucket, there are fossiliferous beds that were recorded by Desor as early as 1849. These fossils have been studied by Wilson (1905), Cushman (1906), and others. The lower beds contain southern species of southern distribution, while those in the upper beds are largely of northern affinity.

Woodworth and Wigglesworth (1934) recognized multiple glaciation in Massachusetts and dated the Sankaty sand from the lower part of the Pleistocene (pre-Manetto of Long Island; table 3). Fuller (1914: 220) places the Sankaty deposits in the second interglacial (Yarmouth). However, MacClintock and Richards (1936) are of the opinion that the lower beds are equivalent to the Gardiners clay of Long Island and date from the last interglacial (Sangamon) while the upper beds correlate with the Jacob sand of slightly younger age.

Hyyppa (1955), on the other hand, dates the Gardiners of Massachusetts from an interstadial within the Wisconsin. He also recognizes a later marine advance which he calls the Taunton interstadial and places between the Tazewell and the Cary stages of the Wisconsin. Hyyppa's report is based upon field work supported by studies of pollen and diatoms which indicate a cool climate.

Fragments of clam shells (*Venus*) and other species occur in the drift near Quincy, Mass., and it has been suggested that these were picked up by the ice from interglacial marine deposits (Nichols and Lord, 1937).

Recent excavations in Boston have shown "evidence for 4—and probably 5—ice advances and 3 marine transgressions" (Kaye, 1961).

CONNECTICUT

The only possible record of a Pleistocene shore line in Connecticut is a deposit of shells which occurs close to low-tide line at Killam's Point on Long Island Sound near Branford, Conn. (Knight, 1933). A fauna of thirty-eight species is recorded including *Littorina irrorata* (Say), a southern species which is sparingly, if at all, known from Long Island Sound today. It is thought that the deposit is of post-Pleistocene age, and represents either the postglacial climatic optimum (Hypsithermal time) which terminated some 4,000 years ago, or possibly only a sheltered cover where *L. irrorata* was able to thrive, although it was not able to obtain a foothold elsewhere in the vicinity.

According to Bloom (1961), the coast of Connecticut has been subsiding for at least 6,800 years and at the rate of $3\frac{1}{2}$ inches per hundred years for the past 1,200 years.

LONG ISLAND AND NEW YORK CITY

The most extensive report on the Pleistocene of Long Island is that of Fuller (1914) who recognized four

TABLE 3

CORRELATION OF PLEISTOCENE OF LONG ISLAND AND NEW JERSEY (SLIGHTLY MODIFIED FROM FULLER AND MACCLINTOCK AND RICHARDS, 1936)

Fuller (1914) Long Island		MacClintock and Richards (1936)	
		Long Island	New Jersey
Wisconsin	WISCONSIN	Wisconsin Harbor Hill Ronkonkoma Manhasset	Wisconsin
Vineyard erosion interval	Sangamon	Jacob sand Gardiners clay	Cape May fm.
Manhasset	ILLINOIAN	— ?	Illinoian
Jacob sand Gardiners clay	Yarmouth	part of Gardiners (?)	Pensauken-Bridgeton Complex
Jameco	KANSAN		
Post-Manetto	Aftonian	Jameco (?)	Jerseyan (?)
Manetto	NEBRASKAN		
		Pliocene ? Manetto	Beacon Hill

FIG. 6. Gardiners clay showing distortion by the Wisconsin ice, Gardiners Island, N. Y.

glacial and three interglacial stages. This interpretation was questioned by MacClintock and Richards (1936) who proposed a somewhat less complicated interpretation, in line with the glacial-control theory, which could be better correlated with New Jersey. The two correlations are given in the accompanying table. Fuller's correlation is very similar to that of Woodworth and Wigglesworth (1934) for Massachusetts (see p. 12).

Both the Gardiners clay and the overlying Jacob sand are well exposed on Gardiners Island where they have been greatly deformed by the overriding ice; fossils have been found in both formations. The fauna has been recorded by Sanderson Smith (1867), Fuller (1914), and MacClintock and Richards (1936). The same relationship of the Gardiners and the Jacob can be seen on Robbins Island and elsewhere. The Gardiners clay contains a fauna rather similar to that prevailing on the region today, while that of the Jacob sand suggests somewhat colder water as demonstrated by the presence of various mollusks including *Neptunea stonei* (Pilsbry), an extinct species frequently associated with deposits of Wisconsin age.

The present interpretation is that the Gardiners clay is of Sangamon interglacial age, while the Jacob sand is of late Sangamon (or very early Wisconsin) age indicating the cooling of the water because of the advancing glaciers.

Shells have been found in wells and shallow excavations in various parts of Long Island and Manhattan. Some of these are probably from the Gardiners clay or the Jacob sand, while others may represent interglacial material that was incorporated in the Wisconsin till and redeposited.

Excavations for the Lincoln Tunnel (formerly called Midtown Tunnel) from 39th Street, Manhattan, to Weehawken, N. J., yielded many mollusks, but they were all species living in the ocean off Long Island and New Jersey and were regarded as postglacial in age (MacClintock and Richards, 1936: 329).

Mollusks found in excavations for a previous tunnel included some species of southern affinity, and these were regarded as interglacial, possibly correlated with the Gardiners (Richards, 1930). The presence of coquina in excavations for the Lincoln Tunnel may also indicate warmer water than that in the Hudson River today even though the species in the coquina were only *Mulinia lateralis* (Say) and *Nassarius trivittata* Say, mollusks common in the area today. On the other hand, these may represent a slightly warmer phase of postglacial time, or may have been reworked from the Gardiners.

Postglacial marine deposits occur north along the Hudson River above New York as far as Storm King where a dwarfed fauna has been reported (Shimer, 1908: 488), which probably represents the maximum advance of the postglacial marine waters north in the

FIG. 7. Jacob sand overlain by Manhassett sand, Robbins Island, Peconic Bay, N. Y.

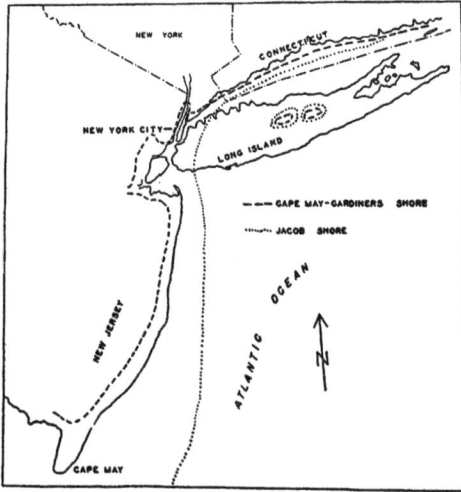

FIG. 8. Approximate position of Cape May (Sangamon) and Jacob (early Wisconsin ?) shore lines (from MacClintock and Richards, 1936).

Hudson Valley. Above Storm King, the Pleistocene clays are devoid of fossils and suggest a fresh-water origin, and in the vicinity of Albany they are definitely varied.

NEW JERSEY

South of the terminal moraine in New Jersey, three Pleistocene formations have been described—the Bridgeton, Pensauken, and Cape May. The first two are non-marine, the third partly marine and partly nonmarine. An older formation—the Beacon Hill—is generally regarded as Pliocene, but may well be early Pleistocene. The large boulders found near Woodmansie and elsewhere may be ice-rafted thus suggesting a glacial dating.

Cooke (1930), Antevs (1929), and others have suggested that the Bridgeton and Pensauken formations might represent high-level shore lines dating from the first two interglacial stages. However, no evidence of a marine origin has been demonstrated. On the other hand, the lithology of the formations—largely sand and gravel—strongly suggests alluvial origin.

MacClintock and Richards (1936) in reviewing the Bridgeton and Pensauken formations give evidence for this alluvial origin and state: "It seems legitimate to consider the older gravels either as one formation with many and complex parts, or as a series of many formations."

It is thought that the Bridgeton-Pensauken complex might represent a fairly long interval of time covering the early and middle Pleistocene. The presence of ice-rafted (?) boulders and the congeliturbation in parts of the Pensauken formation suggests a dating from a

glacial stage, while the warm-climate plants found in Middlesex County, N. J. (Berry and Hawkins, 1935) suggest an interglacial dating.

The Cape May formation was originally regarded as contemporary with the Wisconsin outwash, and thus of glacial age (Salisbury and Knapp, 1917). The presence of a warm-water fauna, especially from Cape May County, cast doubt on the glacial age of the Cape May formation and favored an interglacial dating (Richards, 1933; MacClintock and Richards, 1936). The nonmarine Cape May was traced up the Delaware River where it apparently merged with the Wisconsin outwash somewhere between Camden and Trenton.

Excavations for a canal in Cape May County showed that the formation described as the Cape May was complex (Richards, 1944). The sands and clays of the marine phase were overlain by sands and gravel of fluvial origin. The fauna from the marine Cape May as revealed in the spoil banks contained two distinct faunas: (1) a warm-water fauna thought to be of Sangamon interglacial age, and (2) a cold-water fauna, especially characterized by Neptunea stonei (Pilsbry) thought to be of Wisconsin age.[5]

It was significant that the cold-water fauna was not found in the excavations at Two Mile Beach, only three miles from the canal (Richards, 1933), nor could it be seen where the marine phase is exposed slightly above low-tide line at a few places along the canal.

It may be wondered why the cold-water fauna occurs in the Cape May region since it is generally accepted that the glacial shore line was some ninety miles or more east of the present beach. However, it is believed that the locality might represent an estuary of the Delaware River rather than the open ocean. That the ancient Delaware River cut across the Cape May peninsula is suggested by the unusual thickness of the Cape May formation at the county airport well near Rio Grande, N. J. (Richards, 1945: 896; 1958: 14). Erosion may account for the absence of the cold fauna in certain places.

Newman and Fairbridge (1960) have indicated that the warping of the "Pamlico—Cape May—Gardiners datum from south to north across New Jersey suggests a former marginal bulge, which may be isostatically subsiding, as shown by the anomalous negative behavior of the City of New York tide gauge." The concept of a marinal bulge immediately south of the glaciated area was discussed by Daly (1934: 180) and applied to New Jersey by MacClintock and Richards (1936: 303).

The marine phase of the Cape May formation can be seen above tide today at only two places. One is along the banks of the Cape May Canal at the bridge on the Seashore Road where casts of Mactra sodalissima can

---

[5] Some shells from the Cape May Canal spoil bank were older than 35,000 years by radiocarbon. Personal communication from Wallace Broecker, Lamont Geological Observatory.

Fig. 9. Oyster bed in the Cape May formation on Maurice
River, Port Elizabeth, N. J.

be seen about five feet above low tide in a silt overlain
by a cross-bedded sand. The other locality is along the
Maurice River at Port Elizabeth, where abundant deposits of oysters occur. These shells have been regarded
by some as part of an Indian shell heap, but the presence of double shells as well as very young shells favors
the interpretation of the deposit as a Pleistocene oyster
reef.[6]

Swamp deposits under Philadelphia, Pa., containing
well-preserved remains of the cypress (*Taxodium dis-
tichum*), have been correlated with the Cape May formation and dated by radiocarbon as older than 42,000
years (Richards, 1960: 107), and thus would agree
with an interglacial dating for the Cape May formation.
The "Fish House clay" formerly exposed in pits near
the Delaware River north of Camden, N. J., with its
fresh-water bivalve fauna may also be contemporaneous
with Cape May.

Oceanographic work together with core drilling has
shown a shore line at about 80 fathoms off the coast of
New Jersey (Ewing, Ewing, and Fray, 1960). The
fauna consisted of species indicating a cold, shallow
water, probably representing a glacial stage. The following were included: *Astarte crenata* Gray, *A. undata*
Gould, *Venericardia borealis* Conrad, *Cardium ciliatum*
Fabricius, *Pecten magellanicus* Gmelin, and *Crucibulum
striatum* Say.

A fauna of shallow, cold-water species was described
from a core taken in the Hudson River canyon at a
depth of 1,900 fathoms (3,470 meters or 11,400 feet)
about 180 miles southeast of New York, or 225 miles
east of the mouth of Delaware Bay. It was suggested
that these fossils were carried to this great depth by
turbidity currents. The presence of *Neptunea stonei*
(Pilsbry) suggested a Wisconsin dating (Richards and
Ruhle, 1955).

[6] Beyond the limit of carbon-14 according to an analysis made
at Lamont Geological Observatory.

Bones and teeth of mammals such as the mammoth,
mastodon, tapir, musk-ox, giant moose (*Cervalces*), and
walrus, dredged by clammers thirty to fifty miles off
New Jersey suggest a low sea level of glacial time, and
all except the tapir suggest a colder climate than today.

Veatch and Smith (1939: 44) described the Franklin
shore as a break in the seaward slope of the continental
shelf, off the coast of New Jersey, at about 40 to 55
fathoms. Its age was not determined. A northward tilt
was explained because of incomplete recovery from the
ice load in the north. Recent work has failed to sub-
stantiate the existence of this shore.

DELAWARE

Relatively little field work has been done on the
Pleistocene of Delaware except some early work of
Chester (1884, 1885) on the gravels. The formation
names applied are usually those of New Jersey or
Maryland.

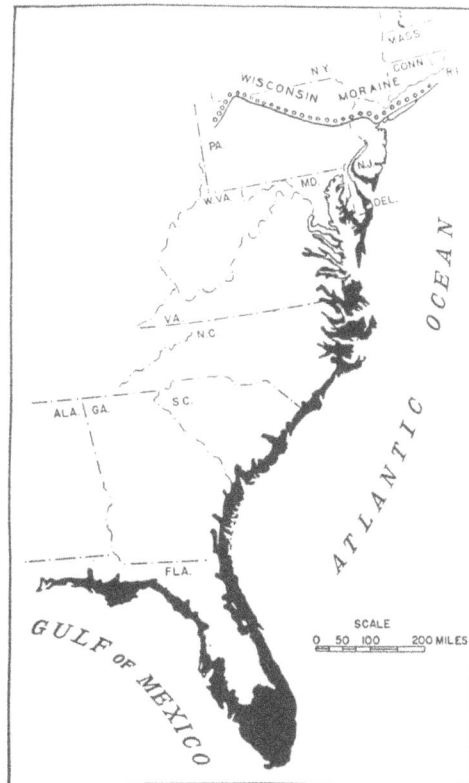

Fig. 10. Pamlico (Sangamon) shore line
(after Richards, 1936).

FIG. 11. Wailes Bluff, Cornfield Harbor, Md.

Coarse gravels found along the Chesapeake and Delaware Canal between St. Georges and the Delaware River resemble those of the nonmarine Cape May of New Jersey.

The localities yielding marine fossils are mostly from shallow excavations and contain few species of significance (Richards, 1936b: 1619). An exception is some oyster shells found at a depth of 8 feet near Laurel (elevation +18 feet).

MARYLAND

Shattuck (1906) described the stratigraphy and fauna of the Pleistocene of Maryland. Of the three formations described, only the lowest, the Talbot, contained marine fossils, and these mostly from two localities in St. Mary's County (localities 15 and 16).[7] These localities are now regarded as part of the Pamlico terrace-formation as redefined by Cooke (1931).

The best-known locality is Wailes Bluff at Cornfield Harbor, near the mouth of the Potomac River, where some seventy-eight species have been listed (Shattuck, 1906; E. R. Smith, 1920; Richards, 1936b; Blake, 1953). It is noted that here, as at Cape May, N. J., the marine phase is overlain by nonfossiliferous sands and gravels, suggesting a shoaling sea toward the beginning of Wisconsin time.

Blake (1953) gave detailed sections at Wailes Bluff and Langleys Bluff (localities 15 and 16) and interpreted the shell bed as of Aftonian age (first interglacial), instead of Sangamon (third interglacial) as suggested by the present writer. The fossils indicate a climate warmer than that prevailing in the region today.

Rasmussen and Slaughter (1955, 1957), in discussing

[7] See Part II for list of localities.

the subsurface correlations of the Eastern Shore of Maryland, attempt the following Pleistocene formations:

WISCONSIN	Parsonsburg sand
Sangamon	Pamlico and Talbot formations
ILLINOIAN	————————
Yarmouth	Walston silt
KANSAN	————————
Aftonian	Beaverdam sand
NEBRASKAN	————————

Marine fossils are recognized only in the Pamlico, although the authors believe that the older formations also are of marine origin.

While no marine fossils have been found in the higher gravels, Cooke (1952, 1958) believes that the shore-line features are not evident because of later dissection. On the other hand, Hack (1955), on the basis of mapping in Prince Georges and Charles counties, believes that the terraces are nonmarine and that there is no evidence of Pleistocene marine transgressions at elevations higher than 100 feet.

VIRGINIA

Clark and Miller (1912) extended Shattuck's formations of Maryland into Virginia. They regarded the Talbot as largely of marine origin, although they realized that the presence of fossil plants at various localities indicated that at least part of the formation was nonmarine. (Again an older marine and a younger nonmarine phase are indicated.)

Wentworth (1930) divided the Talbot of Virginia into three parts as follows:

Princess Anne—Terrace and formation of marine origin. Upper surface ranges from 10 to 15 feet above sea level. Occurs along shores of present tidewater.
Dismal Swamp—Terrace and formation largely of marine origin. Upper surface of marine part ranges from 15 to 25 feet in elevation. Broad area south of Norfolk and elsewhere adjacent to tidewater.
Chowan—Almost wholly fluvial in Virginia. Ranges from 35 feet near the coast to 80–90 feet in elevation near the Fall Belt. In valleys in Chowan River basin and adjacent to other main streams to the north.

Cooke (1931) re-introduced the term Pamlico originally used by Stephenson (1912) for the Dismal Swamp, pointing out its priority, and he also used the term Talbot in a restricted sense for the Chowan. Later, Cooke (1935) doubted the validity of the Princess Anne terrace since it was very local. All the Pleistocene marine localities are now referred to the Pamlico.

Flint (1940, 1942) objected to the correlations of Cooke and presented evidence to show that the higher terraces are of fluvial origin and that only two definite shore lines can be demonstrated in Virginia and the Carolinas. He did not correlate either of these shore lines with any of the terrace names, but used the name "Suffolk scarp" as proposed by Wentworth (1930) for a line approximately equivalent to the inner edge of the

Pamlico terrace. Its toe is between 20 and 30 feet in elevation. The higher shore line, Surry scarp of Wentworth, has an elevation at its toe of 90 feet. The marine origin of the higher scarp is based largely upon physiographic evidence, although, as stated later, there is some fossil evidence in South Carolina.

Moore (1956), in a paper presented before the Southeastern Section of the Geological Society of America in Tallahassee, Florida, described three new Pleistocene formations that underlie the terraces but which are not coextensive with them. However, until his full paper is published, it is impossible to evaluate the correlation and dating of these formations.

The Princess Anne, Dismal Swamp (= Pamlico), and Talbot terraces have been recognized on the Eastern Shore of Virginia (Sinnott and Tibbitts, 1961).

NORTH CAROLINA

Johnson (1907) recognized a series of five well-preserved terraces and remnants of two higher terraces in North Carolina. These were named and mapped by Stephenson (1912). He subdivided the Talbot of Shattuck into two well-marked terraces—the Chowan (60 feet) and the Pamlico (20 feet). According to Stephenson, "the eastern boundary is marked by a well-defined sea-facing escarpment which separates the Chowan terrace from lower-lying Pamlico terrace plain." Cooke extended this subdivision northward into Maryland and Virginia, and southward into South Carolina and Georgia. He substituted the term Talbot for Chowan. In North Carolina marine fossils have been found only in the Pamlico formation.

Some eighteen marine fossil localities were reported by the present writer in 1936 and a few additional ones in 1950. All of these contain a fauna indicating a climate slightly warmer than in the same latitude today. While many of the localities occur close to or below sea level, in other places especially along the Neuse River, they are overlain by nonfossiliferous sands probably equivalent to the upper member at Cape May, N. J., Cornfield Harbor, Md., and elsewhere.

At a point along the Neuse River ten miles below New Bern, N. C., the marine Pamlico overlies truncated cypress stumps, six to eight feet in diameter, which are embedded in dark carbonaceous clay (Mansfield, 1928: 134; Richards, 1936b: 1634). This has been referred to the Horry clay thought to have been deposited during a low sea level in pre-Sangamon time (Cooke, 1937).

As would be expected, little evidence of a Wisconsin low sea-level fauna has been found in North Carolina. However, excavations at Manteo, N. C., in 1946 yielded some cold-water mollusks including *Neptunea stonei* (Pilsbry), a mollusk generally associated with Wisconsin deposits. It is suggested that this locality represents a Wisconsin estuary near Manteo similar to the one reported near Cape May, N. J. (p. 14).

In the swamp country of eastern Virginia and North Carolina, the Pamlico formation is overlain by more than ten feet of peat and plant remains. Dachnowski-Stokes and Wells (1929) have traced the late Pleistocene history of parts of Cartaret County, North Carolina, and show that after the deposition of the Pamlico shells there was a lowering of sea level during which time the peat was deposited. This peat is probably of Wisconsin and Recent age.

Fossils from coquina deposits near Cape Hatteras have recently been discussed by Wells and Richards (1962). They are characterized by species indicating both warmer and colder water, and are regarded as being of Sangamon and younger age. A significant species is the gastropod *Neptunea stonei* (Pilsbry).

SOUTH CAROLINA

The first work on the marine Pleistocene of South Carolina was that of Holmes (1860) who described the rich fauna at Simmons Bluff on Yonges Island, S. C. This fauna was later discussed in some detail by Pugh (1906). The first mapping of the Pleistocene terraces in the state was the work of Cooke (1936). All marine fossil localities cited by Cooke and listed by the present writer in 1936 are referred to the Pamlico terrace-formation.[8]

Excavations for the Santee-Cooper Canal in 1941 revealed a bed of shells at elevation 65 feet, lying above the Eocene limestone. The fauna was a mixture of Pliocene and Pleistocene species. The age is uncertain; it may represent a very late Pliocene deposit, not recognized elsewhere along the East Coast, or it may be an early Pleistocene deposit possibly equivalent to the Penholloway terrace-formation (Richards, 1943a).

Taber, in a paper read before the Southeastern Section of the Geological Society of America at Columbia, S. C., in April 1954,[9] presented evidence for two Pleistocene submergences in South Carolina, an older one at a higher level, for which he used the old name "Lafayette" of early Pleistocene age, and a younger one (Pamlico) dated as Sangamon. Taber would correlate with the older marine transgression the Santee-Cooper bed as well as some diatoms from McBeth, S. C., identified by Dr. E. Hyyppa and quoted by Flint (1940).

Doering (1960) also favors the belief that the "Lafayette" formation represents a marine submergence of the southern Atlantic Coastal Plain in early Pleistocene time. He equates part of the "Lafayette formation" with the Citronelle of the Gulf Coast (p. 21).

Bones of various land animals such as the mammoth, mastodon, bison, horse, and camel have been washed

---

[8] Flint and Deevey (1951: 287–288) report a carbon-14 dating of more than 20,000 years for a cypress stump from the lower part of the Pamlico formation near Myrtle Beach, S. C., thus favoring a Sangamon age. It is possible that the sample came from the underlying Horry clay.

[9] Unpublished.

onto the beaches of South Carolina at Edisto Island and elsewhere. It is thought that these mammals lived during one of the glacial stages when the land extended far beyond the present shore line.

Cooke (1937) described the Horry clay from a section of the Intra-Coastal Canal, 2.5 miles northwest of Myrtle Beach, Horry County, South Carolina. It is exposed also along the Neuse River, 10 miles below New Bern, Craven County, North Carolina. In both places a thin clay deposit lies beneath the Pamlico formation. Near Myrtle Beach the clay contains cypress knees, other plant remains, and numerous diatoms. According to Cooke, "the presence of rooted tree stumps beneath a thick marine deposit that evidently accumulated in quiet water gives conclusive evidence that the sea stood lower on the land when they grew than in the immediately succeeding epoch."

GEORGIA

Veatch and Stephenson (1911) recognized two Pleistocene terraces in Georgia, the Satilla (to 40-foot elevation) and the Okefenokee (to 125-foot elevation). Later work of Cooke (1925, 1943) revealed other terraces, including the Pamlico, to which all marine fossil localities have been referred (Richards, 1936a, 1954).

On the other hand, MacNeil (1950) on physiographic evidence recognized four marine shore lines in Georgia and Florida as follows:

Okefenokee (Sunderland)	150 feet	Yarmouth
Wicomico	100 feet	Sangamon
Pamlico	25–35 feet	Mid-Wisconsin
Silver Bluff	8–10 feet	Post-Wisconsin

The possibility that some of the higher Pleistocene sands of Florida may be residual deposits rather than marine shore lines has been suggested by Altschuler and Young (1960). MacNeil's mid-Wisconsin dating of the Pamlico is questioned by the present writer because there is little evidence of sufficient deglaciation to raise sea level 25 feet above the present during any of the Wisconsin interstadials. The Silver Bluff shore line may be post-Wisconsin, but more likely it is late-interglacial representing a falling sea. This correlation is discussed further under Florida.

FLORIDA

Considerable work has been done on the Pleistocene of Florida, (Matson and Sanford, 1913; Cooke and Mossom, 1929; and others). The present writer still holds, in general, to the correlation published in 1938 (Richards, 1938b), namely that the Anastasia formation, the Miami oolite, Key Largo limestone, and at least most of the Fort Thompson formation can be correlated with the Pamlico formation farther north, and thus should date from the Sangamon interglacial.

On the other hand, Parker and Cooke (1944), recognizing that one of the Pleistocene deposits of southern

Florida—the Fort Thompson—consists of alternating fresh-water and marine limestones, have suggested that the formations represent the various glacial and interglacial stages of the entire Pleistocene. The bands are very thin (only 10 to 20 feet for the entire formation) and the fossils of the various marine and fresh-water zones are repeated. Therefore, it seems better to the present author to regard the Fort Thompson as representing a single interglacial—probably the last. In any case, no shore lines definitely demonstrated by marine fossils are known higher than 30 feet above sea level. This opinion has been expressed also by DuBar (1958).

Parker and Cooke (1944: 286–290) included in the Pamlico formation all marine Pleistocene deposits of Florida younger than the Anastasia. In the region around Lake Okeechobee and the Caloosahatchee River, the Pamlico formation overlies the Fort Thompson. The Pamlico in this area generally consists of barren quartz sand. If the Anastasia and Fort Thompson are correlated with the Pamlico of farther north and are dated as Sangamon, the Pamlico as used in Florida may be equivalent to the barren sands overlying the fossiliferous Pamlico in northern Florida (Rose Bluff on St. Mary's River), in the Carolinas and Maryland, and overlying the marine Cape May in New Jersey. Until detailed field work can be done, it is proposed that this upper barren sand be termed the "upper member," generally of nonmarine origin as contrasted to the Pamlico proper which is marine.

Sellards (1919) has described the Lake Flirt marl as a fresh-water "calcareous mud" along the Caloosahatchee River at Lake Flirt. It underlies the peat deposits of the Everglades in the type area, and overlies the Pamlico sand, or where the Pamlico is missing it overlies the Fort Thompson formation. The maximum thickness of the Lake Flirt marl is about five feet. A characteristic fossil is the fresh-water gastropod *Helisoma scalare* (Jay) (Puri and Vernon, 1959: 239).

Peat containing a cold-water flora of diatoms has been reported from Santa Rosa County. This overlies a de-

FIG. 12. Coquina in the Anastasia formation at "The Rocks," south of St. Augustine, Fla.

TABLE 4

CORRELATION OF PLEISTOCENE OF FLORIDA AFTER
PURI AND VERNON (1939)

Late Wisconsin, Interglacial	Silver Bluff, 8 feet
Late Wisconsin, Glacial	Erosion
Peorian, Interglacial	Pamlico, 30 feet
Early Wisconsin, Glacial	Erosion
Sangamon, Interglacial	Wicomico, 100 feet
Illinoian, Glacial	Erosion
Yarmouth, Interglacial	Okefenokee, 150 feet
Kansan, Glacial	Erosion
Aftonian, Interglacial	Coharie, 220 feet
Nebraskan, Glacial	Erosion. High-level alluvium

posit containing brackish water diatoms which has been correlated with the Pamlico. The cold-water peat has been dated as Wisconsin (Gunter and Ponton, 1933; Hanna, 1933).

Recently DuBar (1958a, b) reinterpreted the Caloosahatchee formation, long regarded as Pliocene on the basis of its extensive marine molluscan fauna, as early Pleistocene. This change was partly based upon the study of some vertebrate fossils, especially *Equus leidyi*. The present writer believes that the evidence of the invertebrates far outweighs that of the vertebrates and prefers to regard the Caloosahatchee formation as Pliocene (Richards, 1959).

Recent investigations by Tertiary workers and collectors in South Florida, including Axel Olsson, have shown conclusively that the richly fossiliferous beds generally known as the "Caloosahatchee" are more closely related, by its fossils and by its stratigraphic position, to the Miocene (Tamiami, of Late Yorktown age), which directly overlies it, than it is to the overlying beds generally assigned to the Pleistocene. This interpretation has been obtained largely through the more critical collecting of its fossil shells, mostly from beds accessible in place, made possible through the extensive excavation work in connection with levee construction and other flood-control operations. The evolutionary sequence of many of our most characteristic Caloosahatchee genera and species can now be traced downward through transitional forms to connect directly with other Miocene shells. Therefore, it is clear that the Caloosahatchee is definitely a part of the Tertiary development along our coastal plain and that any affinities which it may hold with the Pleistocene or with the Recent fauna are hardly more than could be expected from its position near the top of the Tertiary column. It is hoped that much of this newer information will be published in the near future.

There is some evidence of a marine Pleistocene deposit older than the Anastasia-Miami from fossils obtained from a well as Kissimmee in Osceola County, although this latter fauna may be Pliocene (Richards, 1938: 1281–1287).

A series of marine terraces has been recognized in Florida. These are thought to represent "land-marginal marine sediments deposited during cycles of eustatic adjustment in sea level, associated with maxima and minima developments of ice in the Pleistocene" (Puri and Vernon, 1959: 239–240). Five surfaces have been recognized and correlated with high stands of the sea (table 4).

Cooke (1945) has given a somewhat different correlation (table 5).

As stated earlier (p. 18), MacNeil has recognized shore lines up to the Okefenokee (150 feet). On the other hand, Altschuler and Young (1960) have regarded some of the higer Pleistocene sands as residual rather than of margin origin. Specifically, they state that the "quartz sand blanket (in central Florida) is mainly an insoluble residue of the lateritic alteration of the Bone

TABLE 5

PLEISTOCENE OF FLORIDA (AFTER COOKE, 1945: 17)

WISCONSIN	Erosion. Lake Flirt marl (partly recent). Pamlico sand. Shore line at 25 feet. Erosion interval.		
Sangamon	Talbot formation; shore line at 42 feet. Penholoway formation; shore line at 70 feet. Wicomico formation; shore line at 100 feet.	Anastasia formation Miami oolite Key Largo Island	
ILLINOIAN	Erosion interval; fresh-water limestone in the Fort Thompson formation.		Fort Thompson formation (marine and fresh-water)
Yarmouth	Sunderland formation; shore line at 170 feet. Coharie formation; shore line at 215 feet.		
KANSAN	Erosion interval; fresh-water limestone in the Fort Thompson formation.		
Aftonian	Brandywine formation; shore line at 270 feet.		
NEBRASKAN	Erosion interval.		

FIG. 13. "Shore line" at Silver Bluff, south of Miami, Fla.

Valley formation (Pliocene) and not a transgressive Pleistocene deposit."

Altschuler and Young suggest an uplift caused by major faulting in deeper rock to account for the linearity and alignment of the sand ridges.

The sand deposits of "Trail Ridge" in north-central Florida have been regarded as a Pleistocene beach deposit (Cooke, 1939: 42–44). However, in view of the lack of marine fossils, this interpretation must be regarded as unproved. There is some evidence of differential uplift.

As stated earlier, the present interpretation is that the Pamlico terrace-formation should be dated as Sangamon. Sufficient evidence has not been presented to demonstrate intra-Wisconsin stands of the sea higher than present sea level, and certainly not to a height of 30 feet—that of the Pamlico terrace.

It is possible that the Silver Bluff shore line, well exposed just south of the city of Miami, may be intra-Wisconsin, or even date from the Climatic Optimum (Hypsithermal time), as has been suggested. However, according to Price (1956: 160), "the stage of oxidation of the heavy minerals in the Silver Bluff soil and the mapping of the barriers in Florida by MacNeil strongly suggest a Pleistocene age and an origin by the retreating Pamlico sea." [10]

GULF COAST

Price (1956) traced what he considers to be the Ingleside barrier member of the Pamlico terrace from western Florida to Baffin Bay, Texas. The Pamlico is mostly deltaic in Louisiana and Texas, and partly so in Alabama and Mississippi according to Price. [11]

Marine Pleistocene fossils are rare along the Gulf Coast between western Florida and the Mississippi River, although a few localities have been reported, mostly only slightly above sea level, or from below sea level in wells (Richards, 1939a).

Deeply buried shallow-water fossils show that the

sediments carried down the Mississippi River have caused extensive subsidence in the region of the Mississippi Delta. Marine Pleistocene fossils have been reported from wells in Terrebonne Parish, south of New Orleans, to at least a depth of 2,470 feet (Richards, 1939a: 308) and to more than 3,500 feet in a well at South Pass (Askers and Holck, 1957). [12]

The Pleistocene deltaic plains of the Louisiana Gulf Coast were described and named by Fisk (1938, 1944). A correlation with the glacial chronology has been suggested by Fisk and McFarlan (1955). The deltaic formations have been traced downdip toward the Gulf of Mexico where they become marine (Akers and Holck, 1957). The base of the Williana formation, of early Pleistocene age, occurs at a depth of −4,200 feet at the South Pass well near the edge of the continental shelf. A study of the foraminifera in the South Pass well suggests variations in depth of water as would be expected with a cyclic rising and falling sea level of the Pleistocene. Foraminifera in some of the cores indicate tropical conditions associated with coral reefs. These tropical features seem to be related to the marine transgressions of the interglacial stages (Akers and Holck, 1957).

West of the influence of the sinking of the Mississippi Delta, a single Pleistocene marine shore line is known above sea level in Louisiana. Several fossil localities have been reported near Lake Charles, La., at least 20 feet above sea level. The Pamlico shore line has been traced westward from Lake Charles, La., to Beaumont, Texas, where it can be seen at an elevation of between 20 and 25 feet (Price, 1956: 161).

Lagoonal deposits of at least two Pleistocene marine formations have been seen in outcrop on the coast of Texas, and shell beds have been reported from buried Pleistocene strata. Most of the fossils have come from the two younger formations—the Oberlin and Eunice. These were formerly called "Beaumont" (Richards, 1939b). Price has made detailed studies of the Oberlin and Eunice formations in the Corpus Christi area. Each has both deltaic and strand deposits. Oysters and coquina in the Bayside area represent the lagoonal phase of the Oberlin, while similar younger shell deposits represent lagoonal and possibly shallow-water marine deposits of the Eunice along its preserved barrier-and-lagoon terrace, the Ingleside (Price, 1933, 1934, 1958; Price, Cook and others, 1958: geologic column and plate 4).

From Beaumont, Texas, where the Ingleside barrier-and-lagoon terrace is at an elevation of 20 to 25 feet, the terrace can be traced westward to Fannett, Texas, where it "swings seaward diagonally down the slope of the tilted plain to Smith Point on the shore of Galveston Bay." Here the Ingleside terrace "lies at sea level and continues at low elevations between sea level and 10

---

[10] The Silver Bluff shore line may be equivalent to the Epi-Monastirian of the Mediterranean. See page 35.
[11] Personal communication, July, 1961.

[12] The possibility that these fossils may date from more than one interglacial stage has been suggested by Deevey (1950).

feet above sea level and as far southwest as the Rio Grande and Rio Soto la Marina, Tamaulipas" (Price, 1956: 161). According to Price (Price, Cook *et al.*, 1958), the Ingleside barrier and lagoon represents only a part of the Eunice formation of Doering. This barrier correlates, according to his field studies, with at least a part of the Pamlico terrace of Florida as mapped by MacNeil (1950).

Barton (1930) and Price (1933) once described scarps higher than the Ingleside shore line as "terraces," but no marine fossils have been found associated with these features. According to Price (1956: 162) these scarps are now "found to be local tectonic or subaerial erosion features. None of these seem to have originated as, or later become shorelines."

The Citronelle formation, regarded as Pliocene by most writers, is now dated from the early Pleistocene (pre-Nebraskan) on the basis of plant fossils and foraminifera (Doering, 1960). A correlation with the Calabrian of Italy is suggested. In the Gulf Coast area, the Citronelle is thought to be the surface equivalent of the basal member of the subsurface Pleistocene section (= Williana of Fisk).

The Pleistocene of the Gulf Coast, particularly the work of Fisk in the Mississippi Delta, has been summarized in a recent book by Murray (1961).

Late Pleistocene history. Studies on sedimentation and paleontology correlated with radiocarbon dates have suggested the late Pleistocene history of the northern Gulf of Mexico given in table 7 (Curray, 1960; Parker, 1960; Shepard, 1960).

Climatic variations indicated by the fossils are explained more by changes in wind direction, ocean currents, and ecology, than by major changes in temperature.

The fluctuations of the rising sea level of the Texas coast, with correlations to the glacial chronology, are shown in figure 14 (adapted from Curray, 1961: 1708).

Alternate interpretations of the late Wisconsin history of the Gulf of Mexico have been given by other workers.

TABLE 6

CORRELATION OF PLEISTOCENE DEPOSITS OF THE COASTAL PLAIN OF THE GULF OF MEXICO (AFTER FISK, MCFARLAN, DOERING, AND RICHARDS)

Fisk, 1944	Fisk and McFarlan, 1955	Doering, 1956, 1958	Richards, 1939	
Recent "A Series"		Flood Plain Sicily Island		
Prairie	Bradyian Late Tazewell	Holloway Prairie		
Montgomery	Sangamon Late Illinoian	Eunice	BEAUMONT	Ingleside
Bentley	Yarmouth Late Kansan	Oberlin		Upper Beaumont
Williana	Aftonian Late Nebraskan	Lissie	Lissie	
		Citronelle		

TABLE 7

LATE PLEISTOCENE HISTORY OF THE GULF OF MEXICO (AFTER CURRAY, PARKER, AND SHEPARD, 1960)

Years before present

More than 20,000	Beds of oyster and *Rangia* in coquina banks at 10 fathoms off the Texas coast. Radiocarbon dates of 26,900 and 32,500 years B.P. have been determined, but it is thought that these dates may be too young because of carbonate replacement and contamination. These beds may indicate a falling sea.
20,000–18,000	Sea level at minus 65 fathoms or below. A warm fauna explained by changes in circulation of sea water at time of lowered sea level.
18,000–16,000	Sea level rose to about 45 fathoms.
16,000–14,000	Sea level lowered slightly, or was stationary. The presence of *Crassostrea virginica* suggests lagoons.
14,000–12,000	Sea level rose to about 25 fathoms (Two Creeks time).
12,000–10,000	Regression from a high of 22 to 25 fathoms to a low of about 35 fathoms.
10,000–7,000	Sea level continued to rise, although there was a temporary regression between 9,000 and 8,000 years B.P.
7,000 to present	Gradual rise of the sea to present level, but not above.

Broecker, Ewing, and Heezen (1960) favor the concept of an abrupt rise of temperature about 11,000 years ago which they interpret as the end of the Wisconsin glacial stage. While not entirely contradictory to the interpretation of Shepard and Curray, the latter believe that "the 11,000 B.P. date appears to be halfway or more through the major return of water to the oceans. . . . It is more likely that the general warming of the oceans had reached a point where there was a widespread change in the planktonic fauna at that time" (Shepard, 1960: 343).

On the other hand, McFarlan (1961), on the basis of radiocarbon datings from the Mississippi Delta, comes to the conclusion that sea level rose from lower than −440 feet to −200 feet prior to 35,000 years B.P., and

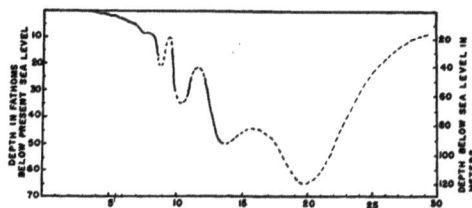

FIG. 14. Fluctuations of sea level in the Gulf of Mexico during the Late Pleistocene (adapted from Curray, 1961).

that there was then a stillstand of the sea to 18,500 years
B.P. Broecker (1961) in discussing McFarlan's paper
disputes the argument for a stillstand, and uses the
same data to support the theory of an abrupt rise in sea
level about 11,000 years B.P.

All workers recognize the fact that studies of sea-level
changes in the Gulf of Mexico in the vicinity of the
Mississippi Delta are complicated by subsidence caused
by the weight of sediments carried down the Mississippi
River. The total length of the Wisconsin stage is probably greater than that shown in Shepard's chart. However, in the "pre-classical Wisconsin might be more than
66,000 years (Dreimanis, 1960), which is a figure close
to estimates of Emiliani (1955: 565) of about 75,000
years and of Flint and Brandtner (1961) of about
70,000 years. Ericson et al. (1961: 345, fig. 24), on
the other hand, give an estimate of 115,000 years for
the total length of the Wisconsin.

WEST COAST OF NORTH AMERICA

SOUTHERN CALIFORNIA

The Pacific coast of California is well known as an
example of a rising coast line. At San Pedro (near Los
Angeles) thirteen terraces have been recognized up to
elevations of 1,300 feet (Woodring, Bramlette, and Kew,
1946). The present location of these terraces or beaches
was formerly explained entirely by deformation of the
land, although more recently it has been recognized that
glacial control of sea level also played a part in the
Pleistocene history of the California coast line. Farther
northwest, near Santa Barbara, seventeen strand lines
have been recognized up to 1,000 feet in elevation (Upson, 1949, 1951).

In the Palos Verdes region (south of Los Angeles)
three outcropping formations are regarded as of early
Pleistocene age. These formations—the Lomita marl,
the Timms Point silt, and the San Pedro sand—are
highly deformed and contain a number of extinct species.
These deformed formations are overlain by a series

FIG. 15. Fossiliferous marine Pleistocene overlying Miocene
at Goleta, California.

TABLE 8

SUGGESTED CORRELATION OF PLEISTOCENE OF SAN PEDRO
DISTRICT (AFTER WOODRING, BRAMLETTE,
AND KEW, 1946)

Formation	Temperature suggested by the fossils	Age
Present sea	61° F.	—
Palos Verdes	63° F.	Sangamon
San Pedro sand	50°–60° F.	Yarmouth
Timms Point silt	50°–52° F.	Kansan
Lomita marl	60°–62° F.	Aftonian

of terrace deposits which are relatively horizontal. The
lowest of these has been named the Palos Verdes formation; the other terraces are unnamed.

Fossils, especially mollusks, are abundant on some of
these elevated beaches, especially on the 25-foot level
near Goleta, Santa Barbara, Newport Bay, and San
Diego. Fossils are also abundant in the deposits of
early Pleistocene age (Arnold, 1903; Grant and Gale,
1931; Woodring, Bramlette, and Kew, 1946; Kanikoff
and Emerson, 1959, etc.). The number of totally extinct
or locally extinct species suggests a pre-Wisconsin age.
Radiocarbon datings as summarized by Bradley (1956)
of more than 39,000 years for shells from the 100-foot
terrace at Santa Cruz, California, support the pre-
Wisconsin dating.

Woodring, Bramlette, and Kew (1946: 100) summarize various tentative correlations for the Pleistocene
deposits in the San Pedro district in table 8.

However, until more is known about the diastrophic
history of the California coast, this correlation must be
regarded merely as a preliminary attempt at dating—
as was intended by the authors.

It is, however, recognized that the fauna of the Timms
Point silt suggests a colder climate than that of the
region today. Since the Palos Verdes—San Pedro
region has been subjected to considerable tectonic movement, it does not seem possible to explain the Pleistocene formations and terraces merely by eustatic control
of sea level.

Emery (1960) has recently reviewed the raised marine terraces of the coast of southern California. He
believes that some of the terraces, if less than 200 feet in
elevation, might possibly be correlated with terraces elsewhere if the rate of Pleistocene diastrophic movement
has been slow enough in relation to changes in sea level.

According to Upson (1951) the shore lines above
200 feet were probably formed during and after
the supposed mid-Pleistocene deformation of southern
California.

Recently five submarine terraces indicating low stages
of the Pleistocene sea have been recognized off southern
California (Emery, 1958, 1960). These are all dated
from the Wisconsin glacial stage by Emery.

Various submerged and emerged terraces have been
recognized along the coast of Santa Rosa Island (about

San Diego District	Palos Verdes Hills	Ventura Basin		Santa Maria District	Kettleman Hills	San Francisco Peninsula	Eel River Basin	Vertebrate Stage	Age
		Northwest Margin	West Central						
Bay Point Fm.	Palos Verdes sand	Unnamed Terraces		Unnamed Terraces				Rancholabrean	Late Pleistocene
	Terraces 2-13							?	
	San Pedro sand / Timms Point silt		Saugus Fm.					Irvingtonian	Early Pleistocene
								?	
	Lomita marl / Santa Barbara Fm.					Merced Fm.	Carlotta Fm.	Blancan	
San Diego Fm.	ss.		Pico Fm.	Careaga sandstone	San Joaquin Fm.		Scotia Bluffs sandstone		Late Pliocene
				Foxen mudstone	Etchegoin Fm.		Rio Dell Fm.	Hemphillian	Medial Pliocene
				Sisquoc Fm.					

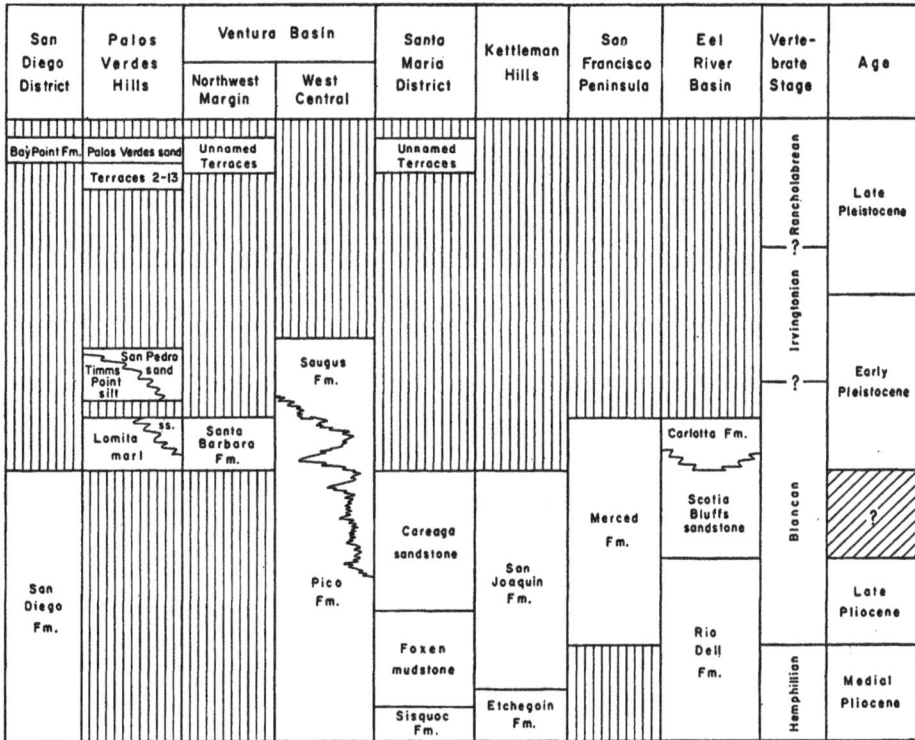

FIG. 16. Tentative correlation of some Pliocene and Pleistocene sections in California (after Valentine, 1961, courtesy Univ. Calif. Press).

forty-five miles southwest of Santa Barbara). The deposits on the lowest terrace, equivalent to the Dume Terrace of Davis (1933), have been called the Santa Rosa Island formation (Orr, 1960), of which three members have been recognized. The two older members (Garañon and Fox) suggest high stands of the sea, and on the basis of radiocarbon and field evidence are dated from the Sangamon interglacial.[18] The youngest member (Tecolete) represents a lowered sea level and is regarded as of Wisconsin age. An area at the midpoint of the sequence was dated by radiocarbon at 29,700 ± 3,000 B.P. (Broecker and Kulp, 1957). The charred remains of the dwarf mammoth are attributed to the activities of man.

Higher terraces on Santa Rosa Island are referred to the Malibu platform with fore edge at about 250 feet, as defined by Davis (1933), from the Santa Monica coast (Orr, 1960: 1115).

Valentine (1961) has recently published an extensive paleoecologic study of the Pleistocene mollusks of Cali-

fornia. Although the purpose of his paper was not to date the various units, he does present a tentative correlation (p. 424) which is reproduced as figure 16 of this report.

Valentine says (p. 426) that, if the Lomita marl is basal Pleistocene, it may be preglacial or

early Nebraskan or even early Kansan, in which case the Timms Point silt and San Pedro sand are most likely of late Nebraskan or late Kansan age. The hiatus between the Early and Late Pleistocene faunas in California may thus include some of either Yarmouth or Aftonian through much of Sangamonian time.

He suggests an early Wisconsin age for the Palos Verdes sand, although, as pointed out earlier in this paper, a Sangamon age must also be considered.

BAJA CALIFORNIA (MEXICO)

The Pleistocene history of Baja California, Mexico, has recently been summarized by Durham and Allison (1960):

Orogenic elevation and depression of the coast line has locally obscured the effects of eustatic sea level changes,

---

[18] Beyond the limit of radiocarbon. Personal communication from Wallace Broecker, Lamont Geological Observatory.

FIG. 17. Twenty-five foot terrace at Punta Descanso,
Baja California, Mexico.

but long stretches of the Baja California Pacific coast appear to have remained static. . . . Anomalously high
Pleistocene coastal terraces occur at various elevations
along the orogenically unstable western side of the Gulf
Trough (C. A. Anderson, 1950: 23, 46–47). Locally, as
near Santa Rosalia (Wilson and Rocha, 1955: 55), marine
terraces have been elevated to at least 340 meters.

Fossiliferous Pleistocene beaches, somewhat deformed, have been recognized along the Pacific coast
of Baja California (Emerson and Addicott, 1958;
Addicott and Emerson, 1960; Beal, 1959; and others).
A carbon-14 dating of a Pleistocene elevated beach at
El Pulmo (lat. 23° 26′ N., long. 109° 25′ W.), at an
elevation of 12 feet was obtained by Olson and Broecker
(1959) as 24,000 ± 1,000 years B.P. Similar datings
have been obtained from other shell deposits in Baja
California (Hubbs, Bien, and Suess, 1960).

The 25-foot shore line is very evident at various places
between the United States border and Ensenada, for
example at Punta Descanso, where a fauna has been
described. The Pleistocene fauna of Baja California,
although indicating minor climatic changes, probably
explained by ecological conditions, contains no extinct
species or species far beyond their present range.

NORTHWEST CALIFORNIA, OREGON AND WASHINGTON

Pleistocene wave-cut terraces have been reported
northwest of San Francisco up to elevation 375 feet
(Weaver, 1949). Similar terraces have been reported
from elsewhere in northwestern California as well as
in Oregon and Washington. The data on these have
recently been summarized by Heusser (1960: 19–24),
but no attempts have been made at dating or correlation.
R. G. Johnson (1962) has made paleoecological studies
of the Pleistocene Millerton formation at Tomales Bay
some forty miles north of San Francisco, but again no
dating was attempted. Marine Pleistocene fossils are
exposed to an elevation of 25 feet near Cape Blanco,
Oregon (Baldwin, 1945). Crustal movements or tectonic activity, other than isostatic adjustment, probably
played an important role in the Pleistocene history of
this area.

BRITISH COLUMBIA

The Pleistocene history of British Columbia, Canada,
involves both changes in elevation caused by the weight
of the ice, and changes in sea level caused by glacial
control. Various fossiliferous deposits, all of late Wisconsin age, have recently been described by F. J. E.
Wagner (1959).

On Vancouver Island, postglacial marine sediments
are known up to 400 feet above sea level, while near the
city of Vancouver they are at 750 feet (Armstrong and
Brown, 1954: 361). In extreme northwestern British
Columbia similar marine deposits occur up to 400 feet
(Hanson, 1934: 181). A recent summary of the late
Pleistocene history of the Pacific Northwest, with spe-
cial reference to the vegetation, has been given by
Heusser (1960).

ALASKA AND WESTERN CANADIAN ARCTIC

Postglacial marine shore lines occur along the southeastern coast of Alaska to elevation 600 feet. In some
places a cold-water fauna has been reported (Twenhofel, 1952).

Pleistocene fossiliferous beaches have long been
known from the vicinity of Nome, Alaska, and the fauna
was described by MacNeil, Mertie, and Pilsbry (1943).
Recently, Hopkins, MacNeil, and Leopold (1960) have
discussed the stratigraphy of these beaches as a "type
section for the Bering Strait region." See table 9.

No evidence of warping or depression of the land
because of the ice was reported.

Schrader (1904) described the Gubik sand along
the Arctic coast of Alaska and regarded it as of Pleistocene age. The formation consists of some 150 feet of
unconsolidated marine and nonmarine sand, and gravel,
silt, and clay overlying the Cretaceous. Marine fossils
have been reported at several places near Point Barrow
(Dall, 1920, and others) but the exact age is uncertain.
In the 1945 glacial map of North America (Flint et al.,
1945) footnote 5 under Alaska states that the marine
deposits along the Arctic coast of Alaska are probably
of interglacial age.

Radiocarbon dates for the Gubik sand recently obtained (Coulter, Hussey, and O'Sullivan, 1960) indicate that the upper member was initiated prior to
38,000 years ago and was terminated prior to 9,100

TABLE 9

PLEISTOCENE OF BERING STRAIT REGION, ALASKA
(AFTER HOPKINS, MACNEIL AND LEOPOLD)

Fourth Beach   elev. 120 ft.   Aftonian
  Iron Creek Glaciation = Nebraskan or Kansan

Third Beach   elev. 75 ft.   Yarmouth
  Nome River Glaciation = Illinoian

Second Beach   elev. 35–40 ft.   Sangamon
  Salmon Lake glaciation = Wisconsin

years ago. This dating is in line with a glacial or inter-glacial dating for part of the formation and precludes a postglacial dating similar to that given to the deposits on the Arctic Islands.

Shell beds probably of post-Wisconsin age are known from Herschel Island (elevation 100 feet) and Kay Point (elevation 25 feet) in Arctic Canada, near the Mackenzie Delta. These are thought to indicate shallow and brackish water rather than true marine conditions as suggested by the Gubik fauna. No elevated beaches were found in the immediate vicinity of the Mackenzie Delta (Richards, 1950). Since the Arctic coast between the Mackenzie River and Bering Strait was unglaciated, it is probable that there was little depression and subsequent uplift caused by the weight of the ice. Furthermore, it is believed that any postglacial rise of the land in the vicinity of the delta would have been counteracted by subsidence caused by sediments brought down the Mackenzie River. The raised beaches at Kay Point and Herschel Island may indicate local uplift.

The Beaufort formation occurs on the western part of the western Arctic Islands from Meighen Island to the south coast of Banks Island (Tozer, 1956; Craig and Fyles, 1961). The age of the formation is probably late Tertiary to early Pleistocene, and thus the formation may be partly equivalent to the lower part of the Gubik sand of Alaska. Driftwood (all beyond the range of carbon-14) and pollen are known from the Beaufort formation, but no marine fossils have been reported.

Peat beds, of probable interglacial age, have been reported from western Banks Island, the Mackenzie Delta, and elsewhere in the western Canadian Arctic, especially in the region not glaciated by the Wisconsin ice (Terasmae, 1959a; Craig and Fyles, 1961).

Elevated beaches containing shells of probable post-Wisconsin age have been found on Victoria Island, and elsewhere in the western Arctic up to about 500 feet above present sea level. Radiocarbon dates of 12,400 years B.P. and later, suggest a late Wisconsin age (Craig and Fyles, 1961).

Farrand and Gajda (1962) have recently plotted the isobases of the marine limit of the Canadian Arctic. The elevated beaches cited by Craig and Fyles lie within the limits of their maps. A few localities, for example on Prince Patrick Island, contain fossil shells higher than the isobases shown on the map, and may indicate an earlier Pleistocene marine invasion, possibly of Sangamon age. Radiocarbon dates of earlier than 38,000 years B.P. on Prince Patrick Island and elsewhere suggest a pre-Wisconsin age.

However, the great majority of the carbon-14 dates from these elevated marine features are of late Wisconsin and postglacial age, leading Farrand and Gajda to the conclusion that the entire Arctic Archipelago was probably glaciated during Wisconsin time.

WEST COAST OF SOUTH AMERICA

COLOMBIA

No marine Pleistocene deposits are known from the Pacific coast of Colombia. The *tablazos* of Ecuador and Peru (described below) dip toward the north and are apparently below sea level along the Colombian coast.

ECUADOR [14]

A flat-lying littoral deposit known as the tablazos, presumed to be Pleistocene, overlies Eocene and Cretaceous sediments along most of the coast of Ecuador. When observed from the sea, the tablazos appear as parallel terraces, and are thought to represent successive uplifts during the Quaternary (Shepard, 1937: 138–139). According to Shepard, the tablazos occur "at roughly 10, 250, and 350 feet above present sea level." However, more recent work suggests that the tablazos are tilted. Furthermore, the tablazos rise toward the south. According to Olsson (1940b: 407) they

appear for the first time on the Galeras Peninsula, at first only a few feet above sea level. They become more extensive further south and at Cabo San Lorenzo and on the Isla de la Plata, the Pleistocene terraces or tablazos occur at several levels as high as 750 feet or more.

Marine fossils are found at many localities, but there is no evidence to indicate changes in climate.[15]

Thus, according to Olsson (1940a: 407), "in contrast with southern Colombia, western Ecuador is a land of recent uplift and arching."

The tablazos are especially conspicuous in the Santa Elena Peninsula where they occur up to elevation 300 feet. Although three main levels can be recognized, there is evidence of considerable local faulting and tilting. The marine fauna is largely of Panamic affinity, and contains no extinct species.[16]

PERU

Like those of western Ecuador, the tablazos of northern Peru form a very conspicuous feature of the topography of the coastal zone. Viewed from the sea, they resemble terraces but actually they are elevated seafloors and represent stands of the sea at different times during the Pleistocene. They extend inland as nearly level plains for many miles, particularly in the Sechura Desert section of Peru. The deposits of the tablazos, consisting of sands, gravel, and shell breccia limestones,

---

[14] The writer is indebted to geologists of the Tennessee del Ecuador and Anglo-Ecuadorian Oilfields, Ltd., for recent information and interpretations on the Pleistocene of Ecuador.
[15] Some of the deposits higher than 100 feet may be pre-Pleistocene.
[16] This fauna is now being studied by R. R. H. Lemon of the Royal Ontario Museum in Toronto. The present writer had the opportunity to make a brief field excursion with Mr. Lemon in the Santa Elena region in June, 1961.

FIG. 18. Horizontal Pleistocene deposits of the Talara tablazo overlying tilted Eocene near Talara, Peru.

vary in thickness; they are comparatively thin in the Talara tablazo (see fig. 18) and much thicker in the Mancora. In parts of the Paita Mountains they may be missing entirely, the tablazo level being shown merely as a planed-off basement floor.

The floor on which the tablazo lies varies according to local geology but in the main it is a nearly level surface, planed-off by the Pleistocene sea as it transgressed over the land, irrespective of the hardness and structure of the underlying rocks. On the Paita Peninsula, much of the basement floor is formed of dense metamorphic and igneous rocks; in other places, the floor may be Cretaceous or Tertiary, predominantly the latter.

The tablazo deposits show evidence of the regional uplift and warping that has affected the coastal area in Pleistocene and post-Pleistocene times. The inter-tablazo intervals were times of general uplift when it may be assumed that the elevated land suffered considerable erosion. From Sechura northward to the Gulf of Guayaquil there is an area of general uplift; south of Sechura to Lima there has been general subsidence.[17]

The tablazos are especially well developed in the vicinity of Talara and have been described by Bosworth (1922) and more recently by Lemon and Churcher (1961).

The following three principal levels have been recognized (Bosworth, 1922):

	Altitude	Thickness
Lobitos tablazo	50–110 ft.	5–15 ft.
Talara tablazo	150–350 ft.	8–25 ft.
Mancora tablazo	200–1200 ft.	75–250 ft.

The oldest tablazo (Mancora) occurs at about 900 feet above sea level at El Alto (Cabo Blanco) and falls steadily to about 300 feet near the Chira River forty-five miles to the south. Coquina and unaltered mollusk

---

[17] The above was partly adapted from a personal communication from A. A. Olsson.

shells are very abundant on the tablazo, and locally (near El Alto) numerous whale remains and shark teeth are known. The fauna suggests relatively cold water, such as that off the coast of Peru today where the Humboldt Current flows northward from the Antarctic.[18]

The middle tablazo (Talara) occurs only as remnants in the Talara area. It drops in elevation from 280 feet near Talara to about 130 feet near the Chira River. The fauna is rather similar to that of the Mancora tablazo.

The Lobitos tablazo is more narrow and probably represents a shorter duration. It falls from about 135 feet above sea level near Mancora to 67 feet at Lobitos. The Lobitos fauna is much richer than those of the older tablazos and contains elements suggesting both warm and cold water. It was suggested by Bosworth and by Lemon and Churcher that this was caused by a meeting of warm and cold faunas near Punta Parinas or Cabo Blanco, as is the case today. During the deposition of the earlier tablazos those promontories did not exist, allowing the cold Humboldt Current to flow close to shore considerably to the north of Talara instead of being deflected seaward. The Lobitos fauna is closely related to that of the tablazos of the Santa Elena Peninsula of Ecuador.

Following the uplift of the Lobitos tablazo, a subsidence enabled the sea to transgress again, destroying much of the Lobitos deposits. This was followed by an uplift which is still in progress. Shell ridges left by this most recent uplift are known as the Salina deposits, and are especially conspicuous near Puerto Chuelo, south of Talara. These shell ridges parallel the present beach and reach over 30 feet in elevation.

The age of the tablazo deposits has not been determined. Suter (1927) and Daly (1934) attempted to correlate the various levels with glacial and interglacial stages. However, the fact that only one or two extinct

FIG. 19. Escarpment between Talara tablazo and Mancora tablazo, south of Talara, Peru.

---

[18] The writer had the opportunity to spend a week studying the tablazos in the Talara area in June, 1961, through the cooperation of the International Petroleum Company, Ltd. Special thanks are due to John D. Tuohy and Fernando Zuñiga of that company for assistance in arranging the field trips.

marine species are present in any of the tablazo deposits plus the fact that the west coast of South America is very active tectonically suggest that these shell beds, certainly those of the Lobitos and Talara tablazo, are relatively young, and date from the late Pleistocene. The Salina shell ridges are probably very recent. It is hoped that radiocarbon datings will be obtained for some of these shell deposits.

An extensive vertebrate fauna has been reported from tar seeps at La Brea, Peru, ten miles southeast of Talara (Lemon and Churcher, 1961). This deposit is regarded as post-Mancora and pre-Talara and suggests a late Pleistocene age.

Marine terraces also occur along the Peruvian coast south of Talara, and are especially well developed at San Juan, Chala, Atico, Ocono, Mollendo, and farther south (Bowman, 1916; Broggi, 1946, and others).[19]

CHILE

The following information has been provided by Dr. Carlos Mordojovich of Empressa Nacional del Petroleo at Tarapaca, Chile:

The northern Chillean coast is a steep fault scarp more than 1000 kilometers long. In most areas the coastal range drops abruptly from an average of 800 meters to sea level, so that in a very few places a sandy shore line is developed; instead a rocky coast line is present.

In a few favorable places a platform is developed, and these places are occupied by the few seaports as Caldera, Antofagasta, and Iquique. According to Dr. Bruggen these platforms are of tectonic origin, and of local significance; nevertheless they give place to true marine terraces with Pleistocene to Recent fossils.

The best example of Pleistocene marine terraces is furnished by the area that ties the Mejillones Peninsula with the coast, north of the city of Antofagasta. The highest place where marine fossils occur is at about 200 meters above sea level.[20]

A discussion of the glacial history of Chile together with data on the shore lines has been given by Brüggen (1950), but no attempt was made to date the marine terraces. Other physiographic data on the shore lines of Chile have been given by Illies (1960).

In summary, the Pacific Coast of South America shows considerable evidence of Pleistocene diastrophism. Thus far it has not been possible to correlate the elevated beaches satisfactorily with glacial events, and no significant changes are indicated by the marine fossils.

EAST COAST OF SOUTH AMERICA

ARGENTINA

There is less evidence of diastrophism along the east coast of South America than along the west coast. Even today, some of the best observations on the Pleis-

tocene shore lines of the east coast of South America, especially in Argentina from Buenos Aires to Patagonia, are those made by Charles Darwin (1846) in his famous voyage of the *Beagle*. He recognized horizontal, elevated beaches at various elevations, especially at about 100 feet, extending for a distance of 1,180 miles. While Darwin attributed these to uplift of the land, they might equally be explained by glacial control of sea level. Darwin's evidence has recently been reviewed and interpreted by Zeuner (1959: 298–300).

Recently Feruglio (1933, 1948, 1950) on the basis of his extensive studies, described Pleistocene marine terraces up to 140 meters (420 feet) in Patagonia. Three interglacial stages were recognized; the oldest and highest carries fauna up to 35 per cent extinct, while some of the younger terraces contain living species including some that are now living farther north, thus indicating a warmer climate.

Feruglio's work is summarized in table 10 adapted from some of his reports.

Auer (1959) has given a detailed interpretation of the postglacial history of Fuego-Patagonia, based largely upon paleobotanical studies. He, like Feruglio, finds evidence of postglacial high stands of the sea. He finds no evidence of warping and is of the opinion that the Patagonian glaciers were too small to cause subsidence and later elevation. He believes, however, that the coast has risen recently and is still rising. This uplift may be associated with the volcanic activity of the west coast of South America.

Very recently studies undertaken by the Lamont Geological Observatory have revealed the presence of a submerged shore line showing a low sea-level stage off the coast of southern Argentina. Cores taken on board the ship *Vema* between 1957 and 1960 have

TABLE 10

MARINE TERRACES OF PATAGONIA, AFTER FERUGLIO

Age	Marine terraces	Molluscan fauna
Postglacial	Terrace of Comodoro Rivadavia	Similar to that of adjacent sea
	a. 6–12 meters	
	b. 10–12 meters	
4th Glaciation	Terrace of Mazarredo 15–30 meters	Living species
3rd Interglacial	Terrace of Bahia Sanguineto y del Escarpado Norte. 30–35 meters	All living Slightly warmer
3rd Glaciation	30–42 meters	
2nd Interglacial	Terrace of Camarones 79–90 meters	All living except Ostrea tehuelche warm water
2nd Glaciation		
1st Interglacial	Terrace of Estancia Cabo Tres Puntas 115–140 meters	35% extinct Slightly warmer
1st Glaciation		
PLIOCENE	131–186 meters	40%–50% extinct

---

[19] Data from Victor Behavides, International Petroleum Company, Ltd., Lima, Peru.
[20] Personal communication.

shown coarse sand with shells at a depth of 480 feet at various places between Bahia de la Plata and Tierra del Fuego (Ewing, Fray, and Dahlberg, 1960).

A study of the fossils—now in progress by the present writer—suggests that they came from shallow, cold water, and favors the view of Ewing *et al.* (1960) for a shore line at about minus 480 feet. Preliminary radiocarbon datings suggest a pre-Wisconsin age, and comparative ice volumes suggest an Illinoian age for some shells (Donn, Farrand, and Ewing, 1962). Others may be of Wisconsin age.

It is difficult to reconcile a glacial shore line at +30 meters (90 feet), as recognized by Feruglio and Auer, with a glacial shore line at −480 feet as interpreted by Ewing *et al.* Perhaps the shore line is tilted seaward, as is the case with some of the tablazos of Peru, although the mollusk fauna suggestive of shallow water does not favor this view. Or, perhaps Feruglio's shore lines are not glacial but date from a falling sea at the end of the last interglacial stage. Or, perhaps, there has been more tectonic movement in the Patagonian region than is generally supposed. It is hoped that detailed collecting together with more radiocarbon studies will help resolve this problem.

FALKLAND ISLANDS

Elevated beaches at 6 meters have been reported from the Falkland Islands. Sub-Antarctic species predominate and suggest a slightly cooler climate than that of today. These, like the low terraces of Patagonia, are regarded as postglacial in age (Adie, 1953).

BRAZIL

Little work has been done on the Pleistocene shore line of Brazil. Bigarella and Freire (1960) report a wave-cut terrace reaching about 13 meters above sea level near Matinho in the State of Paraná. The terrace is dated from "late" glacial time and is correlated with one of the terraces described by Auer from Fuego-Patagonia.

GUIANAS

The term "Corentyne series" is applied to unconsolidated sediments which occur along the coast of the three Guianas (British, French, and Dutch). The following table is adapted from the work of Bleackley and Dujardin (1959).

The Demerara clay formation contains a fauna of Pleistocene to Recent mollusks and merges with the sediments of the Orinoco Delta. It is regarded as recent (post-Wisconsin or post-Würm) and may be equivalent to the Flandrian of the Mediterranean area (p. 36).

The Coropina formation forms a continuous strip of land up to elevation 27 feet, and is tentatively regarded as dating from the last interglacial (Sangamon or Riss/Würm), while the White Sands and Berbice

TABLE 11

PLEISTOCENE OF THE GUIANAS (AFTER BLEACKLEY AND DUJARDIN, 1959)

	Geomorphic unit	Stratigraphic division	Nature of sediment	Age
Corentyne Series	Young Coastal Plain	Superficial deposits Demerara clay formation	Shell beaches etc. Marine fossiliferous clays	Recent
	Old Coastal Plain	Coropina formation	Marine and brackish silty clays and clays	PLEISTO-CENE
	White Sand Plateau	White Sand formation Berbice formation	Sands Sands and clay	

PRE-CAMBRIAN

may date from the previous glacial and interglacial stages respectively.

Preliminary pollen studies (van der Hammen, 1959) from bore holes also give evidence for a transition from the last interglacial through the last glaciation to the Recent.

Nota (1958) in a study of sedimentation in the Western Guiana Shelf observes several low sea-level stages that are correlated with the last glaciation on the basis of carbon-14 dates.

VENEZUELA

Elevated shore lines are known from the coast of Venezuela, but at least some of these are warped, and are thought to have reached their present position because of the tectonic history of the region. Very recently Royo y Gomez (1959) summarized the main events in the Pleistocene history of Venezuela, and reported terraces up to 90 meters in elevation in some places, while on others, no such terraces exist.

The Abisinia formation, in the Cabo Blanco area, (15 kilometers northwest of Caracas), occurs up to elevation 62 meters and is regarded as Pleistocene in age. This area is regarded as highly unstable tectonically (Weisbord, 1957). A full report on the fauna has recently been published (Weisbord, 1962).

Richards (1943b) described fossiliferous deposits up to 15 feet in elevation at Juan Griego on the Island of Margarita which are referred to the Pleistocene.[21] Fossiliferous red to yellow-brown (Falca sands) up to about 70 meters on the island are probably Pliocene or Pleistocene (Taylor, 1960). The pelecypod *Egateria?* sp., suggestive of brackish water, has been identified.

COLOMBIA

Two Pleistocene marine terraces have been described from the island of Tierrabomba (near Cartagena) at

---

[21] The "new species" described by Richards from Juan Griego are now thought to be very close to, if not actually identical with, species now living in Venezuelan waters.

FIG. 20. Pleistocene coral reef, Cozumel Island, Quintana Roo, Mexico.

elevations 20 meters (de Porta and de Porta, 1960). The former was correlated with the Penholloway terrace of the Southern Atlantic Coastal Plain, while the lower terrace was correlated with the Silver Bluff (regression of the Pamlico or Sangamon sea). Extensive paleoecological studies on the faunas were made by the de Portas.

Anderson (1927) had previously reported horizontal Pleistocene marine terraces with corals and mollusks along the Caribbean coast of Colombia to an elevation of 60 feet (20 meters).

CARIBBEAN AREA

It is impossible to generalize regarding the Pleistocene history and shore lines of the islands of the Caribbean. Suffice it to say that former strand lines occur up to elevations of several hundred feet on certain islands, while on other islands no former strand lines can be observed.

The region must be regarded as unstable with diastrophic action—both uplift and subsidence—occurring throughout the Pleistocene and well in to the Recent. Some regions, such as Oriente Province, Cuba, show marked uplift with shore lines up to at least 975 feet, while other islands, such as St. Martins and some of the Virgin Islands, have probably subsided during or after the Pleistocene.

On some of the Caribbean Islands, for example some of the Bahamas, Roatan (off Honduras), and Cozumel (off Yucatan), there are coral reefs or limestone deposits only a few feet—perhaps 3 meters— above present sea level. These give the impression of very recent age and may indicate a slightly higher stand of the sea in postglacial time.

On the other hand, Newell (1960), on the basis of field work in the Bahamas and supported by radiocarbon studies, believes that these reef deposits are of Pleistocene rather than post-Pleistocene age and that the last high stand of the sea in the region was during Sangamon time.

A detailed study of the Pleistocene marine fossils

TABLE 12

PLEISTOCENE OF WEST INDIES (AFTER SCHUCHERT, 1935 AND OTHERS)

Island	Pleistocene deposits	References (other than Schuchert, 1935)
Western Cuba	Marine deposits in Pinar del Rio up to 25 feet 300 foot terraces in Matanzas	Richards, 1935
Eastern Cuba	12 sea terraces in Oriente Province to elevation 975 feet	
Jamaica	Elevated coral rock on benches up to 75 feet; evidence of vulcanism	Trenchmann, 1930
Haiti	Uplifted terraces to 1,460 feet	Woodring, 1924
Dominican Republic	Raised coral reefs from 9 to 130 and probably 325 feet; recent elevation	Vaughan, Cooke, Woodring, et al., 1921
Puerto Rico	Recent uplift and warping; highest terraces about 200 feet; marine fossils at Ponce and Cabo Rojo	Mitchell, 1922 Hubbard, 1923 Kaye, 1959
Grand Cayman	Marine Ironshore formation Island actively rising	Richards, 1955 Matley, 1926
Bahamas	Coral limestone and oolite at low elevations	Newell, 1960
Virgin Islands	Marine Pleistocene limestone only on Anegada	
Dominica	Limestone	
Cozumel (Mexico)	Entire island composed of Pleistocene limestone; up to 30 feet	Richards, 1937
Roatan (Honduras)	Pleistocene coral limestone at West End	Richards, 1938
Curaçao	Bench at 15 to 18 feet; higher bench at 45 to 60 feet	Daly, 1934: 160, 166
Trinidad	Terraces at 20 to 50 feet differential uplift	Liddle, 1946 Renz, 1940
Tobago	Coral rock, slightly terraced to 120 and 158 feet on southwest part of island	Trenchmann, 1934
Barbados	Coral and shell rock to 1100 feet	Trenchmann, 1925, 1933

of some of the Caribbean islands, as well as a survey of the literature, has shown no evidence of major changes of climate within the Pleistocene. Ecological changes must have occurred. For example, the lowering of sea level must have caused many of the islands to have been joined together and to have produced extensive mud flats in areas now beneath the sea. Dr. Robert Robertson,[22] of the Academy of Natural Sciences, believes that the molluscan fauna of the Bahamas was greatly depauperated during the last glacial stage because of these ecological changes, but he can find no evidence of any extensive temperature changes.

Table 12 summarizes data on Pleistocene elevated shore lines on a number of the islands of the Caribbean, but makes no pretense of being complete.

MEXICO AND CENTRAL AMERICA

MEXICO

Relatively little work has been done on the Pleistocene shore lines and marine deposits of Mexico. The Pamlico terrace has been traced southward from Texas into Tamaulipas (Price, 1956) and is briefly discussed on page 21 of this report. The Pleistocene of Baja California is discussed under California (page 23).

Shell beds of late Pleistocene age have been reported about 10 to 20 feet above sea level near Punta Penasco, Sonora, on the east side of the Gulf of California (Hertlein and Emerson, 1956), while terraces at about 20 feet above sea level occur along most of the coast of Sonora from the head of the Gulf of California to Guaymas, as well as on some of the islands in the Gulf (Beal, 1948: 30). On the other hand, Ives (1951) had reported Pleistocene shore lines at 75 and 115 feet and possibly higher along the coast of Sonora. Hertlein and Emerson (1956: 163) state that they did not observe fossiliferous exposures at elevations greater than 25 feet above sea level.

Marine Pleistocene limestone occurs along the coast of Yucatan, and makes up the entire island of Cozumel (page 29).

CENTRAL AMERICA

Marine Pleistocene deposits occur at a few places along the Caribbean coast of Central America. A brief summary, based upon records in the literature, is given in table 13. The deposits at Livingston, Guatemala, regarded as Pleistocene by Powers (1918), are now referred to the Barrios formation of Miocene-Pliocene age overlain by Recent sand and gravel.[23]

Raised beaches, capped with a hard-reef pavement with fragments of coral, occur at about 12 feet and 7 feet above sea level along the coast of British Honduras and are thought to represent Pleistocene or Recent beach deposits (Dixon, 1955: 33).

[22] Personal communication.
[23] Data from Compañía Guatemala California de Petroleo.

TABLE 13
PLEISTOCENE OF EAST COAST OF CENTRAL AMERICA
(AFTER SCHUCHERT, 1935, AND OTHERS)

Country	Pleistocene deposits	Reference (other than Schuchert, 1935)
British Honduras	Raised beaches with coral at 7 and 12 feet	Dixon, 1955
Honduras	Elevated coral reefs on Bay Islands. (See table 12)	Richards, 1938a
Nicaragua	Undeformed marine Pleistocene on east coast; volcanoes active	
Costa Rica	Marine shell beds at Port Limón. (Shows mixture with Pacific forms)	Dall, 1912
Panama	Shell beds up to 10 feet	Vaughan, 1919 Olsson, 1940

Less is known about the marine Pleistocene of the Pacific coast of Central America. Fossiliferous marine Pleistocene deposits are known from low elevations at Puerto Armuellas, Panama (Olsson, 1940a), and elsewhere in Panama and the Canal Zone (Vaughan, 1919). The deposits at Puerto Armuellas reach a thickness of 3,000 feet.

GREENLAND

Elevated beaches have been reported along the east and west coasts of Greenland (Bretz, 1935; Flint, 1948; Laursen, 1950, and references therein). The present writer has identified mollusk shells collected by Dr. Richard F. Flint and Dr. A. L. Washburn from various localities in northeast Greenland, especially from the Mesters Vig area, on the south side of Kong Oscars Fjord. These fossils indicate cold, shallow water, and range in age from 4,960 to 8,780 ±210 years B.P.[24]

Previous work by Nansen (1910) and Böggild (1928) recorded elevated strand lines to 480 feet and questionably 1,800 feet on the west coast and to 1,312 feet on the east coast. (The higher elevations are, perhaps, subject to question.) There is also some evidence of tilting with the higher strand lines toward the north. That some of this uplift may have been caused by epeirogenetic movement unrelated to the ice has been suggested by Daly, 1934: 142) and others.

ICELAND

Opinion differs as to whether Iceland was completely glaciated during the Pleistocene. The oldest marine sediments on the island occur on the north coast where the Tjörnes deposits carry a fauna once thought to be of Pliocene age but now regarded as Pleistocene (Askelsson, 1960b: 28). Other marine shell beds in Iceland are regarded as Early Pleistocene or inter-

[24] Yale Geochrometric Laboratory; unpublished data courtesy of Dr. A. L. Washburn.

glacial. The fauna of at least some localities suggests correlation with the Red-Norwich crag of England of the Calabrian (early Pleistocene). (See Bardason, 1925; Lindal, 1939; and Askelsson, 1960a, 1960b.)

Postglacial uplift of the land has been indicated by Thoroddsen (1905) who recorded the marine limit as follows:

Southern Island around Geysir	110 meters
Reykjanes Peninsula	120 meters
Northwest Land	80 meters

SPITZBERGEN

Studies of elevated beaches and their faunas in West Spitzbergen have been made by Feyling-Hanssen and Jorstad (1950) and Feyling-Hanssen (1955). The results are summarized in table 14 (adapted from Feyling-Hanssen, 1955: 51, 57).

The late-glacial fossils consist only of *Mya truncata* and *Hiatella (Saxicava) arctica*. Those of the postglacial temperate period include *Chlamys islandicus, Mytilus edulis, Littorina saxatilis, M. truncata,* and *H. arctica.* The fauna of the postglacial warm period consists of *Astarte borealis, Heteranomia squamula, Zirfea crispata, Littorina litorea, Astarte elliptica,* and *Cyprina islandica.*

Raised beaches up to more than 100 meters above sea level are well developed in the coastal area of Nordaustlandet (North East Land), Spitzbergen. Radiocarbon dates on driftwood of 4,000 to 7,000 years B.P. suggest that the beach was formed during the Hypsithermal Interval (*Tapes* Sea). Wood at 36 meters and shells at 8 to 44 meters are dated at 9,000 to 10,000 years B.P. Shells which may or may not be *in situ* at 44 to 47 meters are dated between 35,000 and 40,000 years B.P., thus suggesting an extensive ice cover between approximately 10,000 and 35,000 years ago (Blake 1961).

Elevated beaches up to 46.5 meters above sea level are recorded from Brageneset, Nordaustlandet (Donner and West, 1957). Seventeen species of mollusks and brachiopods are listed.

TABLE 14

LATE-GLACIAL AND POSTGLACIAL CHRONOLOGY OF SPITZBERGEN
(AFTER FEYLING-HANSSEN, 1955)

Period	Years ago (B.P.)	Shore line (meters)
Recent and Sub-Recent	2,500–present	−0–3
Postglacial Warm Period	6,500–2,500	3–40
Postglacial Temperate Period	9,000–6,500	40–60
Late-glacial Cold Period	— –9,000	60–90

SCANDINAVIA [25]

PRE-WÜRM

The oldest interglacial marine deposit of Denmark is the Esbjerg beds exposed in southwest Jutland. These beds are thought to lie between the tills of the Elster and Saale glaciations and are correlated with the Mindel/Riss interglacial stage (= Yarmouth). See table 1. The fauna suggests cold water and is characterized by *Yoldia arctica* (Nordman *et al.*, 1928: 88–96). This stage is known as the Holstein interglacial in northern Germany. Fresh-water deposits of this age are also known.

During the last interglacial (Riss/Würm) a sea covered parts of southwest Denmark, adjacent Germany, Poland, Holland, and Belgium. Deposits left by this sea represent the Eemian interglacial, named from the river Eem, a tributary of the Ijsselmeer in Holland. These interglacial deposits are post-Saale and pre-Weichsel (see table 17).

The Eemian fauna of Denmark contains many "Lusitanian" species, in other words is characteristic of southwestern Europe. *Tapes senescens* is an index fossil. Although extinct, this species has southern affinities. Thus the Eem fauna suggests a water temperature milder than that of the present Baltic Sea (Marsden, Nordman, and Hartz, 1908; Nordman, 1928; Nordman *et al.*, 1928).

Not all Danish Eemian deposits are marine. Pollen profiles show a transition from cold to warm to cold (Woldstedt, Rein and Selle, 1951).

The marine Skaerumhede series, known from a deep well about 10 kilometers west of Frederikshavn in Vendsyssel, also probably dates from the last interglacial stage. There were thirty-six arctic species, twenty-two boreal, and twenty-three Lusitanian, so distributed that they reflect a transition from boreal through boreoarctic to high-arctic conditions (Nordman *et al.*, 1928: 101).

Wennberg (1949) regards the Skaerumhede beds as dating from an interstadial within the Würm. Brotzen (1961) correlates these with his Göta Alv interstadial of Sweden, and suggests a correlation with the Farmdalian substage of the United States as defined by Frye and Willman (1960). The Göta Alv interstadial is dated by radiocarbon from between 24,000 and 30,000 years B.P.

POST-WÜRM

The "late-glacial" and postglacial history of the Baltic region has been worked out in considerable detail and is well known in the literature. Both the rise of the land brought about by the release of the ice load and the eustatic rise of sea level are involved. The waters of the Baltic area alternated between fresh and

---

[25] For summaries see Nordman *et al.*, 1928; Wright, 1937; Flint, 1957; and Zeuner, 1959.

FIG. 21. Shell beds at Uddevalla, Sweden.

TABLE 16

LATE QUATERNARY CHRONOLOGY AND SHORE LINES OF
SCANDANAVIA (AFTER HAFSTEN)

	Work of Øyen		Work of Hafsten
Ostrea	22–0 meters		Sub-Boreal and
Trivia	47–22 m.	Sub-Boreal	Atlantic
Tapes	70–46 m.	Atlantic	9–25 meters
Mactra	95–66 m.	Boreal	Boreal
Pholas	142–82 m.		130–75 meters
Littorina	170–130 m.	Younger Dryas	Pre-Boreal
Portlandia	205–170 m.		220–130 meters
Mytilus	220–205 m.	Allerød	

salt. The main stages are known as (1) Baltic Ice Lake, (2) *Yoldia* Sea, (3) *Ancylus* Lake, and (4) *Littorina* Sea, the latter three named for characteristic fossil mollusks in their sediments. The location and extent of these bodies of water are shown in figure 22 (after Fromm, 1953).[26] The term *Tapes* Sea is frequently used for the open-sea equivalent of the *Littorina* Sea. Both *Littorina littorea* and *Tapes decussatus* imply warmer water than does *Yoldia arctica* of the preceding episode. The *Littorina-Tapes* seas are roughly equivalent to Hypsithermal time (see page 12). For approximate time scale of the Scandinavian region, see table 15. The *Littorina* Sea developed into the present Baltic Sea, which had an early *Lymnaea* (fresh-water) stage and a later *Mya* stage. The term Mya period has been used for approximately the last five hundred years beginning with the reintroduction of *Mya arenaria* L. (the common soft-shell clam) into

the Baltic and the northwest coast of Europe (Hessland, 1945).

Similar late Pleistocene marine deposits are known along the coasts of Norway and Sweden as well as on the island of Spitzbergen, and their faunas have been extensively studied by Brøgger (1900–1901), Antevs (1928), Feyling-Hanssen (1955), and others.

In 1956 the present writer collected fossils from the famous shell beds at Uddevalla in Bohuslan in southern Sweden, from various localities near Sarpsborg in adjacent Norway, as well as at Kirkenes in the extreme north of Norway. Many of these species were found

TABLE 15

CORRELATION OF LATE PLEISTOCENE BEDS OF SCANDINAVIA
(ADAPTED FROM SAURAMO)

Climate	Marine stages	Baltic stages	Years
Sub-Atlantic	*Mya*	*Mya*	1000
	Ostrea	*Lymnaea* Lake	0
Sub-Boreal	*Trivia*	*Littorina* Sea	−2000
– – – – – – –	*Tapes*		−4000
Atlantic	*Mactra*	*Mastogloia* Sea	−6000
Boreal	*Zirphaea*	*Ancylus* Lake	
		Echineis Sea	
Pre-Boreal	*Littorina*	Pre-Boreal *Yoldia* Sea	−8000
Younger Dryas		Baltic Ice Lake	
Alleröd	*Yoldia*	Late-glacial	−10,000
Older Dryas		*Yoldia* Sea	−12,000

---

[26] For summaries see Nordman *et al.*, 1928; Wright, 1937; Flint, 1957; Sauramo, 1958.

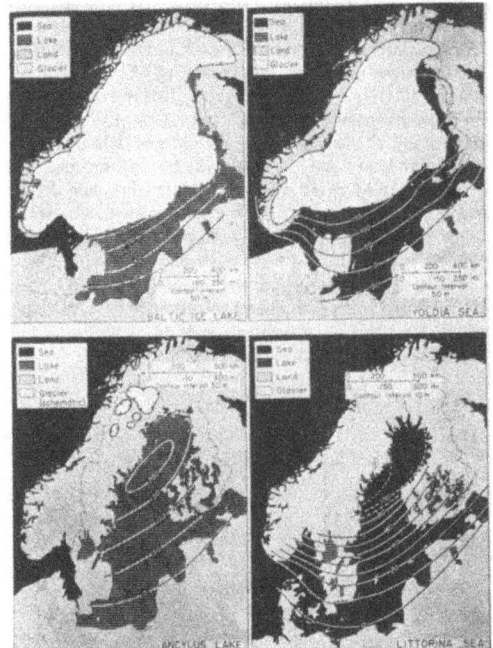

FIG. 22. Successive water bodies of the Baltic Sea basin during the last deglaciation (from Fromm, 1953, as modified by Flint, 1957).

TABLE 17

CORRELATION TABLE OF NORTHERN EUROPEAN PLEISTOCENE. ALPS AFTER ZEUNER (1959); DENMARK AFTER NORDMAN (1928), HANSEN AND NIELSON (1960); GERMANY AFTER WOLSTEDT (1955); NETHERLANDS AFTER ZONNEVELD (1959) AND ZAGWIJN (1960, 1961); ENGLAND AFTER VAN DER VLERK (1955) AND WEST (1955). CORRELATIONS HIGHLY TENTATIVE BELOW GÜNZ/MINDEL INTERGLACIAL

Alps	Denmark	North Germany	Netherlands	East England
WÜRM	MORAINES C-H	WEICHSEL	WEICHSELIAN	"Last Glaciation"
Riss/Würm	Skaerumhede series Eem deposits	Eem Interglacial	Eemian interglacial	Ipswichian interglacial
RISS	MORAINE B	SAALE	SAALIEN	Gipping glacial
Mindel/Riss	Upper Esbjerg *Tellina* clay Lower Esbjerg	Holstein interglacial	Needian interglacial	Hoxnian interglacial
MINDEL	MORAINE A	ELSTER	ELSTERIAN	Lowestoft glaciation
Günz/Mindel		Cromer interglacial	Cromerian interglacial	Cromerian interglacial
GÜNZ ?		Weybourne glacial ?	Menapian ?	ICENIAN
Donau/Günz ?		Tegelen interglacial ?	Waalian ? Eburonian ?	Weybourne crag Chilesford beds
DONAU ?		Brachter glacial ?	Tiglian ? Praetiglian ?	Norwich crag Red Crag

to be identical with those collected in eastern Canada and the northeastern part of the United States.

Table 15 shows an attempt at correlations of the late Pleistocene beds of Scandinavia largely adapted from Sauramo (1958: 44).

Hafsten (1958) has studied the late Quaternary history of the Olsofjord region of Norway on the basis of pollen, and attempted to correlate his data with those based upon studies of the shell beds. Table 16 is adapted from Hafsten (1958: 87) and shows his interpretation of the late Quaternary chronology compared with the work of Øyen.

NETHERLANDS

Three Pleistocene marine invasions of the Netherlands have been recognized from a study of borings, and their faunas have been listed by Van der Vlerk and Florschültz (1953) and others. The Eemian, dating from the last interglacial, is the best known, and its fauna has been studied by many writers, most recently by Van Voorthuysen (1958). A slightly higher water temperature is indicated.

An older marine invasion (Needian, Holstein, or Stör) is dated from the Mindel/Riss interglacial. These deposits are known from borings in Schleswig-Holstein. The fauna is poorer than that of the Eemian, and consists of fewer than sixty marine species (Van der Heide, 1957: 272).

There is also a marine deposit dating from the Early Pleistocene. The Amstelian-Icenian represents a continuous marine series indicating a climate gradually becoming colder.

A more detailed chronology of the Pleistocene of the Netherlands has been based upon pollen studies of Zonneveld (1959), Zagwijn (1960, 1961), and other workers. Prior to the Elster (= Mindel), three warm stages are recognized (Cromerian, Waalian, and Tiglian[27]), and three cold stages (Menapian, Euborian, and Praetiglian). Whether these be actual interglacial and glacial stages, or merely times of warmer and cooler climate, has not been determined. See table 17. At least part of the Icenian marine deposits interfinger with those of the Tiglian.

Tectonic movements, mostly the downward warping of the North Sea and the sinking of the Rhine Delta, have caused most of the interglacial deposits of the Netherlands to be below present sea level. See Brouwer (1956) and summaries by Charlesworth (1957: 1265–1266).

GREAT BRITAIN

There is no "standard" terminology for the glacial chronology in the British Isles. That used in table 17 (from West, 1955) applies to East Anglia where the best marine sequence is exposed. There is evidence of three definite interglacials (Ipswichian, Hoxnian, and Cromerian). Furthermore, the cold faunas of some of the crags may possibly indicate one or more earlier glaciations. The existence of glacial high sea levels may possibly be explained by a general lowering

[27] There is some confusion about the terms Tiglian and Tegelen. According to van der Vlerk (1953: 40), "the clay of Tegelen was formed during a period which is named Tiglian (Tiglia is the Latin name of Tegelen)."

FIG. 23.   Pleistocene Red Crag overlying London clay (Eocene)
at Bawdsley, England.

of sea level throughout the Pleistocene on which has
been superimposed the oscillations caused by glacial
control (see page 7).

Present opinion, as expressed by a Commission at
the 18th International Geological Congress in London
in 1948, is to place the Pliocene-Pleistocene boundary
at the base of Red Crag of East Anglia, which is
equivalent to the Calabrian and Villafranchian of the
Continent. This might assume a fairly long period
of time before the "first"glaciation as previously recognized (= Günz).

An attempt has been made to determine the number
of non-extinct species of marine mollusks in the various
Pleistocene deposits with the following results (Baden-
Powell, 1948).

Weybourne crag	89%	
Corton beds	88%	Lowestoft glacial or Cromerian interglacial
Butley crag	87%	part of Red Crag
Newbourne crag	68%	middle part of Red Crag
Walton crag	64%	lower part of Red Crag
Coralline crag	62%	Pliocene

The East Anglia section has been correlated with that
of the Netherlands (Van der Vlerk, 1955). See table
17.

An illustrated discussion of the crag faunas has
recently been given by Chatwin (1961).

It has been suggested that some of the older Pleistocene deposits of eastern England might represent
deltaic deposits of an ancient Rhine River.

The correlation of the Pleistocene deposits of other
parts of Great Britain has been summarized by Flint
(1957: 402–406), Zeuner (1959: 133–172), and
Wright (1937), although the latter's Pliocene-Pleistocene contact is much too low according to present
correlations.

Mitchell (1960) has recently summarized the Pleistocene history of the Irish Sea and western England.

TABLE 18
PLEISTOCENE OF THE IRISH SEA REGION (AFTER MITCHELL, 1960).
SEE ORIGINAL PAPER FOR NUMEROUS SUBDIVISIONS

	Time unit	Sea level
UPPER PLEISTOCENE	Recent warm period (postglacial)	rising
	Smestow cold period	low
	Ipswich warm period (interglacial)	as today
	Gipping cold period	low
MIDDLE PLEISTOCENE	Hoxne warm period	plus 100 feet
	Loowestoft cold period	low
	Cromer warm period	plus 20 feet
LOWER PLEISTOCENE	Red crag and higher crags	varied; up to plus 200 feet

He does not give correlations with the Continent, but
instead divides the Pleistocene into Lower, Middle,
and Upper. His time and sea level are summarized in
table 18. See table 17 for correlation with elsewhere
in Europe.

Zeuner (1948, 1959) has recognized various high
sea levels and has attempted to correlate them with
those of the Mediterranean area in the Thames River
Basin; high sea levels have been reported at "600 ft.,"
"400 ft.," 200 ft., 100 ft., 60 ft. and "a few feet" above
tide (Zeuner, 1959: 292). The levels above 200 feet
are questionable.

Probably the best ancient strand lines of the British
Isles can be seen on the Island of Jersey in the English
Channel where they have been recognized at 7.5, 18,
33, and 57 meters (Mourant, 1933, 1935; Zeuner,
1959).

Postglacial marine deposits are known up to an
elevation of 100 feet in parts of Scotland. These contain an arctic marine fauna and were probably laid
down while glacial ice was not far distant. There is
also a lower strand line, at about 30 feet, carrying a
warm fauna which is probably postglacial in age. This
lower strand line occurs in parts of Scotland, the north
of England, and in northern Ireland (Wright, 1937;
Flint, 1957: 255–256; Donner, 1959). These deposits
are similar to late- and postglacial deposits of Scandinavia and northeastern North America, and indicate an
upwarping of the land as it recovered from the load
of the glacial ice. The 100-foot shore line is tilted,
but the lower ones are horizontal (Donner, 1959).

Postglacial beaches are known from the east coast
of Ireland, but there is no positive evidence of an
earlier Pleistocene marine transgression (Stephens,
1957).

MEDITERRANEAN AREA

In parts of the Mediterranean area elevated strand
lines and associated fossils have been correlated with
all three of the major interglacial stages, and analogous
shores are found in many other parts of the world.

TABLE 19

CORRELATION OF PLEISTOCENE MEDITERRANEAN SHORE LINES (ADAPTED FROM NUMEROUS SOURCES)

Altimetric	Paleontologic	Stratigraphic division	Average elevation	Correlation	Marine fauna
Postglacial	Flandrian		+3 meters to −3 meters	Holocene	
				Würm	
Epi-Monastirian	"Tyrrhenian III"	1st Interstadial	3–4 meters		warm *Strombus* fauna
Late Monastirian	Tyrrhenian II	* —?—	7.5 meters	⇐ Eemian	
Main Monastirian		Last Interglacial	17.5 meters		
				Riss	
Tyrrhenian	Tyrrhenian I	Great Interglacial	30 meters 40 meters?		*Strombus* fauna
				Mindel	
Milazzian	Sicilian	First Interglacial	60 meters		cold
				Günz?	
Sicilian		Villafranchian	100 meters		cold
				Donau?	
Calabrian	Calabrian		180 meters		cold

* There are two main possibilities about the Epi-Monastirian terrace. It may represent the last oscillation of the Eem (as most Dutch geologists say); or it may be the first interstadial of the Würm (as Zeuner and some of the German geologists say).

Some writers have expressed the opinion that the land has been stable and that the beaches represent the actual position of sea level at various stages of the Pleistocene. However, at the present time, it is generally recognized that there has been considerable tectonic movement, at least in parts of the Mediterranean basin, causing some of the shore lines to be displaced from their original position.

There is not complete agreement on the correlation of the Mediterranean strand lines, or even on the terminology. To complicate matters, the same name has been used for altimetric levels as well as for faunal units, even though they may not be of the same age. Table 19, adapted in part from Zeuner (1959), gives the correlations used today by many students of the Mediterranean Pleistocene.

The data on the Mediterranean strand lines have been discussed by many authors (see recent summary by Zeuner, 1959). The pioneer studies are those of Lamothe (1911) on the coast of Algeria, Deperet (1906) on the Italo-French Riviera, Dubois (1924) on the north coast of France, Gignoux (1913) in Italy and Sicily, and Blanc (1937) in Italy.

The Tyrrhenian fauna definitely suggests warmer water than that of the Mediterranean today and is characterized by the gastropod *Strombus bubonius* and other species typical of Senegal on the west coast of Africa. As noted in table 19, this fauna is correlated with several shore lines and, according to Zeuner, probably dates from the last two interglacial stages.

The Sicilian and Calabrian faunas suggest cooler water and are characterized by the pelecypods *Cyprina islandica* L. and *Mya truncata* L. and the gastropod *Buccinum undatum* L. Subdivisions of the Sicilian and

Calabrian have been recognized by Ruggieri and Selli (1948) and others.

The Epi-Monastirian shore line at 3 to 4 meters, dated by Zeuner (1959: 378–379) from the first interstadial of the Würm glaciation, is widespread. It may be confused by local uplift; it may be obscured by local subsidence, or, indeed, it may belong to the last interglacial, the opinion held by Fairbridge (1958) and others. Current opinion is that sea level was not appreciably higher than at present during any of the interstadials of the last glaciation (see discussion of the Silver Bluff shore line of Florida on page 20).

The term Flandrian was proposed by Dubois (1924) for the transgression following the last glacial low sea level. This stage was recognized along the coast of Belgium and France. A similar stage in Italy has been called the Versillean. The culmination of this ("The Climatic Optimum") is equivalent in time to the *Littorina* stage of the Baltic. Flandrian deposits occur up to 3 meters above sea level, although in places they are only known from below present sea level.

The Tyrrhenian shore lines are probably the best developed of the various shore lines and can be traced along the south shore of the Mediterranean from Turkey to Morocco and along the north coast in Italy, France, Spain, and Portugal. Recent work of Gigout, Selli, and Blanc [28] has suggested that in places these shore lines, as well as the older ones, are warped and do not represent the true position of sea level of the various stages of the Pleistocene. On the other hand, according to Gigout,

[28] As discussed at a symposium on the Mediterranean held at Burg Wartenstein, Austria, in the summer of 1960, and sponsored by the Wenner-Gren Foundation for Anthropological Research. To be published in *Quaternaria* in Rome.

the Atlantic Coast of Morocco has been very stable. Gigout (1956) recognized the following shore lines in this area.

Calabrian	100 meters
Sicilian	60 meters
Tyrrhenian	30 meters
unnamed	20 meters = Monastirian?
Oulgien	5 meters
Versillian or Flandrian regression	
Dunkerquian	2 meters

The Moroccan faunas have been studied by Lecointre (1952).

Bonifay and Mars (1959) have suggested the abandonment of the terms Monastirian and Milazzian and a correlation as shown in table 20.

The *Strombus* fauna is mostly characteristic of the Eutyrrhenian.

On the other hand, Castany and Ottman (1957) believe that there is only one Tyrrhenian shore line. These authors think that the occurrence of various deposits containing the *Strombus* fauna can be explained by differential uplift of the land. "When these deposits are in place, they are always found at low elevations, always near sea level." Thus the *Strombus* or Tyrrhenian shore line sometimes occurs above sea level and sometimes below.

According to Castany and Ottman, the deposition of the Tyrrhenian or *Strombus* beds was followed by the Grimaldian regression, which in turn was followed by the Flandrian transgression. Thus, by inference, the Tyrrhenian would date from the last interglacial (Riss/Würm). Castany and Ottman also call attention to the fact that the cold (*Cyprina*) faunas of the Calabrian and Sicilian always occur in the regions of unstable tectonics. They believe that the cold faunas can be explained because the shells lived in deep water and were subsequently uplifted. This is also the conclusion of Emiliani, Mayeda, and Selli (1961) who studied the isotopic (O^{18}) temperatures of these faunas.

POLAND

Although several interglacial stages have been recognized in Poland (Halicki, 1950; Ruhle *et al.*, 1961),

TABLE 20

SUBDIVISIONS OF THE TYRRHENIAN
(AFTER BONIFAY AND MARS)

Present altitude (feet)	Original elevation of beach (feet)		Terminology (Bonifay and Mars)	Former name	Age
0–15	2–3	T Y R R H E N I A N	Neotyrrhenian	Monastirian	Würm interstadial
0–30	6–10		Eutyrrhenian	Tyrrhenian	Russ/Würm interglacial
20–100	20–25		Paleotyrrhenian	Milazzian	Mindel/Riss interglacial

marine beds are known only from the latest (Masovian II interglacial = Eemian). The geological map of Poland (Ruhle *et al.*, 1961) shows marine interglacial deposits near Elblag (east of Gdansk), and Brodniewicz (1960) has recorded some Eemian mollusks from a boring in Brachlewo, on the lower Vistula, while Mrozek (Galon *et al.*, 1961: 81–84) records an outcrop with an Eemian fauna at Gniew on the Vistula.

Traces of the *Littorina* shore can be seen along the Baltic coast of Poland (Rosa *in* Galon *et al.*, 1961: 64).

U.S.S.R.

INTERGLACIAL

Deposits correlated with the last interglacial stage (Riss/Würm) have been reported from various places in northern European Russia, and have been referred to as the "Boreal Transgression." Mollusks indicating a climate slightly warmer than that of today have been reported from the east coast of the Kola Peninsula facing the Barents Sea, north of latitude 68° (cited by Gromov, 1945; Flint, 1957: 400). Other records of marine interglacial deposits in European Russia include those at Irinowa near Lake Ladoga (Yakolev, 1923), bluffs along the Mga River (Yanichevsky, 1923), various parts of Karelia (Sauramo, 1929), and the Kanin Peninsula (Ramsay, 1904).

Marine interglacial deposits correlated with the Eemian (last interglacial) have been reported from Latvia and Estonia (Dreimanis, 1947, 1949).

The Boreal Transgression is also rather extensively developed in Siberia (Sachs and Strelkov, 1959). The late Quaternary [29] glaciations of Siberia were largely limited to highland areas while much of the Arctic coastal region was unglaciated especially east of the Lena River. Therefore, there was little, if any, crustal warping and later elevation caused by the weight of the ice, and in most places the latest marine invasion on northern Asiatic Russia was the Boreal Transgression of the last interglacial stage. Table 21 shows the glacial chronology of Siberia as recognized by Sachs and Strelkov (1959) and together with suggested correlations proposed by Farrand (1961).

The duration of the Boreal Transgression is thought to have been 40,000 years, occurring between 105,000 years B.P. and 65,000 years B.P. as determined from a study of bottom sediments in the Arctic Ocean (Belov and Lapina, 1958).

Less is known regarding the earlier Quaternary marine deposits of the U.S.S.R. Deposits containing *Glycymeris, yessoensis, Cardita pultunensis,* and other species on the northern coast of the Sea of Okhotsk may date from the early part of the Anthropogene, while a marine sand in the Chuckchi Peninsula with *Cardium ciliatum* is dated from the middle Anthropogene. Some

---

[29] Many Soviet geologists use the term "Anthropogene" instead of Quaternary.

TABLE 21

SIBERIAN GLACIAL CHRONOLOGY (AFTER SACHS AND STRELKOV, 1959) AND SUGGESTED CORRELATION WITH NORTHERN EUROPE, THE ALPS AND NORTH AMERICA (AFTER FARRAND, 1961)

		Siberia	N. Europe	Alps	North America
RE-CENT		POSTGLACIAL	POSTGLACIAL		
UPPER	PLEISTOCENE	Spartansky glaciation Karginsky interstadial Zyryansky glaciation Kasantsevky marine deposits	Weichsel Salpauselka Alleröd	Würm	Wisconsin Valders Two Creeks
		Boreal Transgression	Eemian	Riss/Würm	Sangamon
LOWER MIDDLE	PLEISTOCENE	Maximum Glaciation Interglacial	Saale Holstein	Riss Mindel/Riss	Illinoian Yarmouth
LOWER		Glaciation Pre-glacial	Elster	Mindel Günz	Kansan Nebraskan

marine deposits in the Yugor Peninsula of the Siberian Plain are believed to be correlated with the Holstein (second) interglacial transgression. These deposits contain species such as *Littorina saxatilis* (?) and

Cardium edule, which are absent from the deposits of the younger Boreal Transgression (Sachs and Strelkov, 1961). Most of these deposits are included in the "Northern Transgression" of Lavrova and Troitsky (1960) which preceded the Boreal Transgression, but was more limited.

POSTGLACIAL

A postglacial history similar to that of the Baltic and Scandinavia took place in parts of Arctic European Russia. Shell beds correlated with the *Littorina* Sea have been reported from Petsamo (formerly in Finland) by Sauramo (1929) and in the Kola Peninsula (Lavrova, 1960). Postglacial elevated beaches have been reported from Novaya Zemlya, Franz Josef Land, and the Byiaranga Mountains of Siberia (Koettlitz, 1898; Ourvantzev, 1930). The beaches apparently occur at decreasing elevations toward the east, probably because of the local and alpine nature of the glaciation east of the Lena River or tectonic uplift of the Ural mountain chain which terminates northward in Novaya Zemlya.

Recent studies on the Quaternary of the U.S.S.R. include those of Gerasimov and Markov (1939), Sachs and Strelkov (1959, 1961), and Lavrova (1960).

TABLE 22

SCHEME OF RELATIONS BETWEEN QUATERNARY MARINE DEPOSITS OF THE CASPIAN, BLACK SEA, AND MEDITERRANEAN (AFTER P. V. FEDEROV, 1961)

Mediterranean (H. Alimen, 1955; F. Zeuner, 1959)		Black Sea (Level of the old terraces in the Caucasus)	Caspian Sea		
Recent beds		Nymphean terrace (1–2 m.)	New Caspian stage	Late New Caspian terrace (−23 m.)	
Interval (?)		Thanagoriisk interval (−2−−4 m.)		Derbentian interval	
				Maximum New Caspian terrace (−22 m.)	
Flandrian (3–4 m.)		New Black Sea terrace (4–5 m.)		Chelekenian interval	
				Early New Caspian beds (−24 m.)	
Interval		Old Black Sea beds (−5−−10 m.)		Mangyshlak continental suite (interval −50 m.)	
		New Euxinian beds (−10–40 m.)	Khvalynian stage	Upper Khvalynian horizon (terraces: −16 m., −11 m., −2 m.)	
Epi-Monastirian (2–4 m.)		(?)		Interval −50 m.	
Interval		Epi-Karangatian continental suite (extensive interval)		Lower Khvalynian horizon (terraces: 25 m., 35 m., 48 m.)	
Tyrrhenian II	Late Monastirian (8–10 m.)	Late Karangatian terrace (12–14 m.)		Atelian horizon	Continental suite
	Main Monastirian (18–20 m.)	Early Karangatian terrace (22–25 m.)	Khazarian stage		Marine ("Girkanian") beds
Interval		Interval		Extensive interval	
				Upper Khazarian horizon	
Tyrrhenian I (30–40 m.)		Euxino-Uzunlarian terrace (35–40 m.)		Lower Khazarian horizon (terraces in the Caucasus 85 m., 125 m., 160 m.)	
Milazzian (60 m.)		Old Euxinian terrace (60 m.)		Urundzhikian horizon (terrace in the Caucasus 200 m.)	
Sicilian (100 m.)		Interval	Bakinian stage	Upper Bakinian horizon (terrace in the Caucasus 250 m.)	
		Chaudian-Bakinian terrace (100 m.)			
		Lower Chaudian horizon		Lower Bakinian horizon	
Calabrian (180 m.)		Tanaisian (Khaprovian) continental suite		Tiurkian continental suite	
		Gurian stage (Upper Pliocene)		Apsheronian stage (Upper Pliocene)	

Level of the Caspian Sea—28 m.

BLACK SEA

The Pleistocene history of the Black Sea, although related to that of the Mediterranean, also involved changes in water level due to flooding by meltwaters from the Don, Volga, and other rivers. Connections between the Black Sea and the Caspian Sea and the Sea of Aral occurred also at various times during the Pleistocene (see summaries by Ramsay, 1930: 38–41, and Daly, 1934: 201–204).

A recent attempt at correlation between the terraces of the Black Sea and those of the Mediterranean has been made by Federov (1961). Table 22 shows the subdivisions of Quaternary of the Black and Caspian seas from Federov (1961: 103) with minor changes and translation provided by Professor Federov (personal communication). Changes in level of the Black and Caspian seas during the Holocene have been discussed by Federov and Skiba (1960). The late Quaternary fauna of the Black and Caspian Sea region has been discussed by Nevesskaya (1958) and Nevesskaya and Nevesskii (1961).

IV. CORRELATION BETWEEN EASTERN NORTH AMERICA AND THE MEDITERRANEAN

The question arises as to why three interglacial stages are represented in the Mediterranean area with beaches up to several hundred feet above sea level, while in eastern North America there is paleontological evidence of only a single interglacial high stand of the sea with physiographic evidence of possibly one additional shore line. No answer is available that is entirely satisfactory. However, let us consider several possibilities.

1. One possible answer is that the older shore lines and fossils have been obliterated by weathering and erosion. But why has this been so much greater in America than in Europe?

2. Another possible answer lies in movement of the land. It is generally agreed that the higher shore lines of the Mediterranean area have been considerably uplifted and deformed, with perhaps less warping of the lower beaches. Contrary to the situation along the Mediterranean, the Coastal Plain of the southeastern United States is thought to have been relatively stable during the middle and late Pleistocene time. It is obvious that direct correlation of shore lines on the two sides of the Atlantic entirely on the basis of elevation, as has been attempted by several workers, has no validity in view of the different disastrophic histories of the two regions.

It is possible that further research will demonstrate that the Atlantic Coastal Plain has been subjected to Pleistocene deformation, possibly of a block or non-warping variety. Hack (1955: 38) states that, in view of the recognized Pleistocene deformation of parts of the Gulf Coastal Plain, it is difficult to believe that the At-

lantic Coastal Plain could have remained stable. It is just as plausible to think that the Atlantic Coastal Plain subsided as to think that the shores of the Mediterranean rose. While it is true that the Mediterranean area is part of a very mobile belt, it is also possible that the Atlantic Coastal Plain may represent part of a subsiding geosyncline.

Thus, it is possible that the older interglacial shore lines lie below present sea level along the Atlantic Coast owing to later subsidence of the land, and are not observable today because they lie beyond the present shore line. Low Pleistocene shore lines of the Mediterranean have been recognized by Blanc (1937).

3. It is generally supposed that the older interglacial shore lines are higher than those of the last interglacial. But this may not necessarily be the case. In Italy and Lebanon the older terraces may be very high because of the later elevation of the land. Sea level itself may have been at about the same elevation during the three interglacials. Thus, the deposits of the three interglacials may be superimposed on each other along the Atlantic Coastal Plain. A possible objection to this theory is the presence of raised beaches on the Island of Jersey and on the Atlantic Coast of Morocco, regions that are thought to have been relatively stable.

4. As has been pointed out by many writers, and recently summarized by Zeuner (1959: 205), the fluctuations caused by glacial control of sea level seem to have been superimposed on some major cause which has depressed sea level throughout the Pleistocene and probably since early Tertiary time and possibly earlier. Deglaciation alone cannot explain the highest sea levels attributed to early Pleistocene shore lines.

It has been estimated that during Tertiary time sea level was at least 300 meters higher than at present, and that it has been gradually falling. If the fluctuations caused by glacial control were superimposed upon this slowly falling sea, it might well be that during the "low sea-level stage" of the first glaciation, sea level was higher than it is today. This might explain the cold *Cyprina* fauna in the elevated beaches of the Calabrian and Sicilian as well as the cold, glacial(?) crag faunas of England. This is an alternate explanation to the theory of uplift from great depths of the sea proposed by Castany and Ottmann (1957).

The cause of this general fall of sea level has not been determined. It is possible that the bottom of the sea may have subsided several hundred meters since the beginning of the Tertiary. Umbgrove (1939: 125) says that "some deep ocean-basins must in all probability have originated in a recent geologic part."

5. There has probably also been considerable elevation of the land since the beginning of the Tertiary. However, the position of the ancient strand lines cannot be explained only as a result of tectonic movements. The similarity of levels in many parts of the world make this explanation unacceptable, for it is unreasonable to

suppose that the land rose equally in such different parts of the world as the Mediterranean, South Africa, and the east coast of South America.

6. One reason for the lack of correlation between shore lines in different parts of the world is the lack of uniformity in the method of recording the exact levels of the shore lines being compared. This point is well brought out by Zeuner (1959). Sometimes high-water mark is recorded, while other times it is the low-water mark that is recorded as the shore line. In fact, some of the localities being compared do not represent shore lines at all, but rather are ancient sea floors. Until complete agreement on the method of comparing levels has been achieved, there will be some error in long-distance correlations.

V. POSTGLACIAL CHANGES OF SEA LEVEL

The term Flandrian was proposed by DuBois (1924) for the postglacial (Holocene) transgression which occurred in northern France and Belgium. The deposits were subdivided as follows:

1. Lower Flandrian (Ostend beds) −15 to −30 meters; transgressive with slight regression at the end. 12,000 to 7,500 years B.P.[30]
2. Middle Flandrian (Calais beds) 0 to −15 meters; transgressive with slight regression at the end. 7,500 to 3,500 years B.P.
3. Upper Flandrian (Dunkirk beds) about present sea level; transgressive. 3,500 years B.P. to present.

Minor oscillations in sea level have been noted during the deposition of the Flandrian marine sediments with the result that some occur below sea level and some slightly above (to about 3 meters). A summary of these oscillations has been given by Fairbridge (1958) who has worked out a chronology largely on the basis of radiocarbon studies. Buried peat deposits at Pelham Bay, N. Y., off Florida, Louisiana, the Bahamas, and elsewhere, as well as youthful-looking drowned reefs along the west coast of Florida, are dated from slightly lower stands of the sea, whereas emerged coral reefs in Australia, New Zealand, Algeria, and many other places indicate a slightly higher Flandrian sea.

Newell (1960), on the other hand, believes that the sea level is now near its highest position since the Pleistocene and suggests that the elevated terraces of some Pacific islands frequently cited as effects of recent high sea level, should be reexamined. Jelgersma and Pannekoek (1960) also find no evidence of sea levels higher than that at present in the Netherlands.

VI. SUMMARY

1. *Eastern Canadian Arctic.* Elevated shore lines up to nearly 1,000 feet above present sea level are well

documented. These are all regarded as post-Wisconsin in age and represent uplift following the release of the land from the weight of the ice. The term "Tyrrell Sea" has been used for this embayment in the James Bay area.

2. *Labrador and Newfoundland.* A similar post-Wisconsin sea covered parts of Labrador and Newfoundland. It is believed that the northwest part of Newfoundland is emerging while the southeast part is submerging.

3. *St. Lawrence lowlands.* The late Pleistocene sea invaded the St. Lawrence lowlands to a point one hundred miles west of Montreal as well as north along the Saguenay, north beyond Ottawa, and south into the Lake Champlain lowlands. This embayment has been termed the "Champlain Sea." No evidence of a pre-Wisconsin marine Pleistocene invasion has been demonstrated.

4. *Northeastern New England.* Postglacial marine shells have been found along the coast of Maine up to an elevation of 300 feet. Apparently the sea covered coastal Maine in late Wisconsin time when part of the meltwaters were returned to the sea and before the land had fully recovered from the weight of the ice. This is thought to have taken place about 11,000 years ago. Outwash from a late Wisconsin ice advance overlies the marine deposits (Presumpscot formation) in southwestern Maine. There is some evidence of a very recent submergence of the land that may still be taking place.

5. *Massachusetts.* No trace of a late or post-Wisconsin invasion by the sea has been recognized south of Boston. Shell beds on Nantucket Island (Sankaty) are probably of Sangamon age.

6. *Long Island.* The Gardiners clay is probably of Sangamon age. It crops out in eastern Long Island and especially on Gardiners Island where it has been greatly deformed by the ice. It is below sea level in western Long Island. The Gardiners clay is overlain by the Jacob sand with its slightly cooler fauna; this is regarded as transitional to Wisconsin time. Postglacial silts from beneath the Hudson River carry a fauna similar to that of today. These extend up the Hudson River to Storm King, N. Y.

7. *New Jersey.* The lower part of the Cape May formation is regarded as the time equivalent of the Gardiners clay and is dated as Sangamon, largely on the basis of its warm fauna. While the marine fossils generally occur below sea level, they are found above sea level at a few places in southern New Jersey, notably at Port Elizabeth and Cape May. Sands containing a cold fauna overlie the marine Cape May formation at Cape May, and probably indicate an estuary of the Delaware River during Wisconsin time.

8. *South Atlantic states.* There is paleontologic indication of a warm interglacial sea from Delaware to Florida. Marine fossils are found up to an elevation of

---

[30] Radiocarbon dates after Fairbridge (1958).

28 feet in the Pamlico formation which is dated as Sangamon. Physiographic evidence of a higher shore line (Surry scarp) is also indicated. The only paleontological evidence of an earlier Pleistocene shore line comes from South Carolina (Santee-Cooper Canal), but this deposit may be Pliocene.

9. *Florida.* The interglacial shore line can be traced along both coasts of Florida. The Anastasia formation, the Miami oölite, the Key Largo limestone, and at least part of the Fort Thompson formation, are regarded as being of Sangamon age. Evidence for older Pleistocene marine deposits are not clearly demonstrated.

10. *Gulf Coast.* The Ingleside barrier member of the Pamlico has been traced westward along the coast of the Gulf of Mexico. The Pamlico formation is mostly deltaic in Louisiana and Texas, and partly so in Alabama and Mississippi. The marine Pleistocene is deeply buried in southern Louisiana because of subsidence caused by the weight of the sediments carried down by the Mississippi River. The Pleistocene deltaic plains of the Louisiana Gulf Coast have been described by Fisk, and a suggested correlation given with the glacial time scale has been given by Fisk and McFarlan. One, and possibly two, Pleistocene shore lines have been traced along the coast of Texas and south into Mexico.

11. *California.* The position of the shore lines along the Pacific Coast of the United States has been determined by both the tectonic history and glacial control of sea level. To date, no completely satisfactory correlation has been worked out. In the Los Angeles area, the Lomita marl, Timms Point silt, and San Pedro sand are deformed and are regarded as "Early Pleistocene," while the Palos Verdes formation and the many coastal terraces are regarded as "Late Pleistocene." Several of these shore lines extend south into Baja California, Mexico. The 25-foot terrace is especially conspicuous and is regarded as Late Pleistocene. Well marked shore lines are noted on some of the offshore islands, notably Santa Rosa.

12. *Alaska.* Post-Wisconsin shore lines have been reported from the coast of British Columbia and southeastern Alaska up to an elevation of 750 feet. Their history is similar to that of the shore lines of the eastern Canadian Arctic and the St. Lawrence lowlands. The region near Nome is regarded as stable, and the three shore lines and deposits are dated from the three interglacial stages. The Gubik sand of the Point Barrow area is probably partly interglacial. Post-Wisconsin shell beds are known from Herschel Island and Kay Point in Arctic Canada, a short distance east of the international boundary.

13. *Western South America.* A series of elevated sea floors (tablazos) occurs along the coast of Ecuador and northwestern Peru, and are marked by extensive faunas. Although there is little if any evidence of faulting, these tablazos are definitely tilted. The maximum elevation is about 1,200 feet near El Alto (Cabo

Blanco), Peru. These deposits have not been definitely dated, although the composition of the fauna (all living species) suggests the Late Pleistocene. Terraces and shore lines occur locally along the southern coast of Peru and in Chile.

14. *Eastern South America.* A series of terraces occurs along the extreme southern coast of Argentina (Patagonia) and these have been correlated with various episodes of the Pleistocene. Recent borings taken off the coast of southern Argentina have revealed the existence of an old shore line at −480 feet. If this shore line is of glacial age, as suggested by the fauna and preliminary carbon-14 studies, it is difficult to correlate it with the supposed glacial shore line at +90 feet as indicated by Feruglio. Elevated shore lines occur also along the coast of Brazil, the Guianas, Venezuela, and the north coast of Colombia.

15. *Caribbean Area.* The Caribbean area is very unstable tectonically. Some portions have undergone considerable uplift, while others have subsided or remained stable. Numerous Pleistocene shore lines are known, but these have not been correlated with the glacial sequence.

16. *Greenland.* Elevated beaches are known along both east and west coasts of Greenland, indicating early post-Wisconsin submergence and emergence.

17. *Iceland.* Interglacial Pleistocene deposits are known from various places in Iceland. At least some (Tjörnes) are regarded as Early Pleistocene, possibly correlated with the Red Crag of England and the Calabrian of Italy.

18. *Scandinavia.* Evidence for two interglacial marine submergences are known from Denmark (Esbjerg beds = Yarmouth and Eem = Sangamon). The postglacial history of Scandanavia has included uplift of the land following the release of the load of the ice, as well as eustatic control of sea level. The main events in the Baltic area are as follows: (1) Baltic Ice Lake, (2) *Yoldia* Sea, (3) *Ancylus* Lake, and (4) *Littorina* Sea.

19. *Netherlands, Great Britain, and Northern Germany.* Several marine interglacial submergences are known from these areas, the best known being that of the Eem invasion (last interglacial). Tectonic movements, mostly the downwarping of the North Sea Basin, have caused most of the interglacial deposits of the Netherlands to be below sea level. The Red Crag deposits of East Anglia, England, with their cold fauna, may date from an early Pleistocene glaciation. It may be above sea level today because of tectonic movement, or because of a general lowering of sea level which has taken place throughout the Pleistocene and on which has been superimposed the variations of sea level caused by glacial control. Late or postglacial marine deposits and shore lines are known from northern England and Scotland.

20. *Mediterranean Area.* Strand lines along both shores of the Mediterranean have been correlated with

all three of the major interglacial stages. However, an exact correlation is impossible at this time partly because of the lack of knowledge of the tectonic history of the region, and partly because of a confusion in terminology. It is now generally believed that many of the Mediterranean shore lines, and their shell deposits, have been displaced tectonically, and that the present position of the beaches does not necessarily represent the sea level at the time the shells were deposited. The late Pleistocene deposits with their warm Tyrrhenian fauna are probably the best developed, and at least part of these date from the last interglacial. The Sicilian and Calabrian beaches, with cold faunas, are probably of early Pleistocene age. They may owe their present elevated position to later uplift, or to a general decline of sea level as was suggested for the Red Crag deposits of England.

21. *U.S.S.R.* The Boreal Transgression which covered parts of northern European Russia and western Siberia is dated from the last interglacial stage. These deposits are especially well developed in the Kola Peninsula. The Boreal Transgression was probably contemporaneous with the Eem sea of Holland, Denmark, etc. There is also evidence for earlier Quaternary marine invasions in Asiatic Russia. Late or postglacial marine deposits are also known from the Kola Peninsula and parts of the Soviet Arctic. The fauna and history are similar to those described from Scandinavia, especially northern Norway.

22. Opinion differs as to whether there has been any stand of the sea higher than at present after the end of Wisconsin time. Minor fluctuations have been recognized at various places in Europe, the Americas, and elsewhere. On the other hand, others believe that the last higher stand of the sea occurred in Sangamon time.

PART II

THE MARINE PLEISTOCENE MOLLUSKS OF EASTERN NORTH AMERICA

VII. IMPORTANT PLEISTOCENE FOSSIL LOCALITIES BETWEEN HUDSON BAY AND GEORGIA

LABRADOR
(after Packard, 1865, and Dawson, 1873)

1. "Labrador."
2. Caribou Island.
3. Mouth of Salmon River.
4. Mouth of Sandwich Bay.
5. Hopedale.
6. Pitts Arm, head of Cheatam Bay.

HUDSON BAY
(after Richards, 1940, 1941, and Nichols, 1936)

1. Churchill, Manitoba. Gravel pit five miles south of town, along spur of Hudson Bay Railroad. Elevation 80 feet.[1]
2. Eskimo Point, N.W.T. Low ridges of sand, pebbles, and shells about 150 yards from shore. Elevation about 6 feet above high tide.
3. Five miles south of Sir Bibby Island. About sixty miles north of Eskimo Point, N.W.T. Shallow excavations.
4. Term Point, N.W.T. Thirty-seven miles south of Cape Jones at southern end of Ranken Inlet. Shallow excavations up to 30 feet above high tide.
5. Chesterfield Inlet, N.W.T. Raised sandy beach, one mile south of Royal Canadian Mounted Police Barracks; a few feet above tide.
6. Entrance to Baker Lake, N.W.T. Raised beach at the "Narrows" at east entrance to Baker Lake, near Police Barracks, about two hundred miles northwest of Chesterfield Inlet. Elevation 50 feet.
7. 1¼ miles north of Baker Lake Post. Elevated beach at west end of lake (second gravel ridge from modern beach). Elevation 100 feet.
8. Belcher Islands. Various localities. (Richards, 1940).
9. Sugluk, Quebec. Elevation 224 feet (Nichols).
10. Wolstenholme, Quebec. Elevation 548 feet (Nichols).
11. Dundas Harbor, Devon Island. Elevation 145 feet (Nichols).
12. Wakeham Bay, Quebec. Elevation 110 feet (Nichols).
13. Port Harrison, Quebec. Elevation 200 feet (Nichols).

14. Coral Harbor, Southampton Island, N.W.T. Elevation 134 feet (Nichols).
15. Craig Harbor, Ellesmere Island, N.W.T. Elevation 10 feet (Nichols).
16. Ponds Inlet, Baffin Island, N.W.T. Elevation 65, 60, 120 feet (Nichols).
17. Pangnirtung, Baffin Island, N.W.T. Elevation 0 to about 10 feet (Nichols).

JAMES BAY
(after Richards, 1936)

1. Moosonee, Ontario. Low bluffs along Moose River.
2. Charlton Island, N.W.T. Beach wash.
3. Cary Island, N.W.T. Beach wash.

FIG. 24. Sketch map of Eastern Canada showing marine Pleistocene fossil localities.

---

[1] Localities 1 through 7 after Richards (1941).

42

4. Shipsands Island, mouth of Moose River, Ont.
5. Charles Island, Moose River, Ont.
6. Mouth of Butler Creek, Moose River, north of Moosonee, Ont.
7. Stag Island, on Rupert River, twenty-five miles from mouth.

FIG. 25. Sketch map of Newfoundland showing marine Pleistocene fossil localities.

NEWFOUNDLAND

(after Richards, 1940)[2]

WEST COAST—NORTH TO SOUTH

1. Castor River, fifteen miles north of Hawk Bay.
2. Narrows, River of Ponds Lake.
3. Portland Creek Pond, southwest feeder.
4. Near mouth of Portland Creek. Northeast side of river.
5. Parsons Pond, eastern shore.
6. Winterhouse Brook, Bonne Bay, near mouth.
7. Mouth of Shoal Brook. South arm of Bonne Bay.
8. South end, St. Joseph's (Hell) Cove. South arm, Bonne Bay.
9. Brough's Brook, Bonne Bay.
10. East side of mouth of main river, east of Lomand, Bonne Bay.
11. Woody Point, Bonne Bay, near Whitehouse Creek.
12. Near Paynes Cover, Bonne Bay. Elevation 102 feet.
13. Trout River.
14. North Head, between Bay of Islands and Cape St. Gregory.

[2] Most of the fossils were collected by Richard F. Flint, Paul MacClintock, Helgi Johnson, A. K. Snelgrove, J. Wheeler, and others. See Richards, 1940.

15. Blow Me Down Creek, Bay of Islands, one-quarter mile from shore.
16. Lark Harbor. 35-foot terrace, shells from about 15 feet.
17. Lark Harbor.
18. York Harbor. Clay bank along eastern shore. Elevation 8 feet.
19. Shoal Point.
20. Corner Brook, near railroad station.
21. Corner Brook, terrace about 250 feet back from shore and about 40 feet high.
22. Lower Frenchman Head, Bay of Islands.
23. 1000 feet west of locality 22.
24. Petipas Cove, Humber Cove, Bay of Islands.
25. Mouth of Stone Brook, north arm, Bay of Islands.
26. Mouth of Liverpool Brook, north arm, Bay of Islands.
27. Southeast Point, Tweed Island, Bay of Islands.
28. Crow Head, St. George Peninsula.
29. West Bay Landing, Port au Port Bay.
30. Cove 600 feet south of locality 29.
31. Southwest end, Young's Cove, St. George Bay.
32. One-half mile south of Barrachois Brook, St. George Bay.
33. Four-tenths of a mile south of Harbor Point, north of Highland Church, St. George Bay.
34. One mile west of Stevenville Church. Elevation 35 feet.
35. Robinson Head. Broken shells in outwash glacial sand. 100 feet above tide.
36. Contact between delta and moraine between Middle Barachars Brook and Crabbes Brook, about one mile north of Jeffries.
37. Codroy (Road). Fragments, 15 feet above tide.

NORTH COAST

38. Bay Verte.
39. Brook, one mile west of Clam Pond, southern arm, Little Bay area, Notre Dame Bay.
40. Tilt Cove, Notre Dame Bay. Elevation 60 feet.
41. Springdale, White Bay. Red marine silts.
42. Botwood. 25-foot terrace on west side of Northern Arm Point.

NEW BRUNSWICK

(largely after Packard, 1865, and Dawson, 1873)

1. Lawlors Lake (Packard, Dawson).
2. Duck Cove (Packard, Dawson).
3. St. John.
4. Red Bluff, near St. John.
5. Lancaster (Packard).
6. Black Point.
7. Manarragonis.
8. Campobello Island (Packard).

FIG. 26. Sketch map of Maine showing marine
Pleistocene fossil localities.

MAINE

(after Dawson, 1873; Clapp, 1907; Bloom, 1960; and
material in various museums)

1. Bangor, Penobscot County.
2. Portland, Cumberland County.
3. Saco, York County.
4. Casco Bay, Cumberland County.
5. South Berwick, York County.
6. Augusta, Kennebec County.
7. North Haven, Knox County.
8. Lewiston, Androscoggin County.
9. Scarboro, Cumberland County.
10. Eastport, Cumberland County.
11. Zebbs Cove, Cape Elizabeth, Cumberland County.
12. Yarmouth, Cumberland County.
13. Kennebunk, York County.
14. Wells Beach, York County.
15. Gardiner, Kennebec County.
16. Brunswick, Cumberland County.
17. Calais, Washington County.
18. Waterville, Kennebec County.
19. Freeport, Cumberland County.
20. Gray, Cumberland County
21. Buxton, Cumberland County.

NEW HAMPSHIRE

(after Dawson, 1873)

1. Portsmouth, Rockingham County.

ST. LAWRENCE GULF AND RIVER, QUEBEC

1. Trois Pistoles.
2. Tadoussac.
3. Saguenay.
4. Rivière du Loup.

5. Murray Bay.
6. Beauport.
7. Quebec.
8. Matane River.
9. Montreal.
10. Clarenceville.
11. St. Nicholas.
12. Upton.
13. Acton.

ONTARIO

1. Perth.
2. Pakenham Mills, Cornwall.
3. Green Creek, Ottawa.
4. Ottawa.

PRINCE EDWARD ISLAND

1. Campbellton.

MASSACHUSETTS

1. Sankaty Head, Nantucket County.
2. Quincy, Norfolk County.
3. Point Shirley.

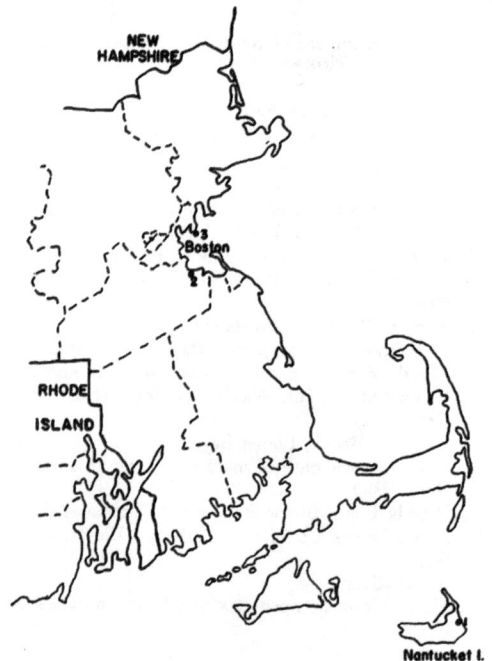

FIG. 27. Sketch map of Massachusetts showing marine
Pleistocene fossil localities.

VERMONT

(largely from Howell and Richards, 1937)

1. Alburgh, Grand Isle County.
2. Isle La Motte, Grand Isle County.
3. South Hero, Grand Isle County.
4. Grand Isle, Grand Isle County.
5. Swanton, Franklin County.
6. St. Albans, Franklin County.
7. Colchester, Chittenden County.
8. Winoski, Chittenden County.
9. Mallets Bay, Chittenden County.
10. Charlotte, Chittenden County.
11. Vergennes, Addison County.
12. Panton, Addison County.
13. Chimney Point, Addison County.
14. Burlington, Chittenden County.

CONNECTICUT

1. Killams Point, Branford, New Haven County.

NEW YORK

LONG ISLAND

(after MacClintock and Richards, 1936).

1. Brooklyn, Kings County.
2. Jones Beach, Nassau County.
3. Gardiners Island, Suffolk County. South end of cliffs, south of Tobacco Lot Bay.
4. Southampton, Suffolk County. Six miles east.
5. Westhampton, Suffolk County. Well between depths of 116 and 137 feet.
6. Robbins Island, Suffolk County. Bluff along south shore of island in Peconic Bay. Jacob sand overlying Gardiners clay.
7. Fire Island, Suffolk County. Well between 0 and 54 feet.

MANHATTAN

8. New York, N. Y. Various excavations in lower Manhattan.

9 Hudson River Tunnels. Excavations for Midtown (= Lincoln) Tunnel as well as material from older Pennsylvania Railroad tunnel. Largely post-Wisconsin.

ST. LAWRENCE RIVER

10. Ogdensburg, St. Lawrence County.
10a. Excavations for St. Lawrence Seaway, St. Lawrence County.

LAKE CHAMPLAIN

(after Goldring, 1922)

11. Mooers, Clinton County.
12. Freydenburg's Mills, Saranac River, Clinton County.
13. Plattsburg, Clinton County.
14. Laphams Corner, Clinton County.
15. Valcour Island, Clinton County.
16. Port Kent, Essex County.
17. Willsboro, Essex County.
18. Port Henry, Essex County (a few miles north).
19. Crown Point, Essex County.

NEW JERSEY

(after Richards, 1933; MacClintock and Richards, 1936)

1. Sandy Hook, Monmouth County. Well between depths of 0 and 100 feet.
2. Union Beach, Monmouth County. Former beach along Raritan Bay near Keyport. Elevation about 10 feet.
*3. Perth Amboy, Middlesex County. Hydraulic fill for Victory Bridge.
4. Sayreville, Middlesex County. Borehole between depths of 38 and 42 feet.
*5. Belmar, Monmouth County. Hydraulic fill on south side Shark River.
*6. Lavalette, Ocean County. Hydraulic fill.
*7. Beach Arlington, Ocean County. Hydraulic fill on north side of Manahawkin Road.

---

* These fossils came from hydraulic fills and shallow excavations, and may be contaminated with recent specimens.

Fig. 28. Sketch map of Long Island showing marine Pleistocene fossil localities.

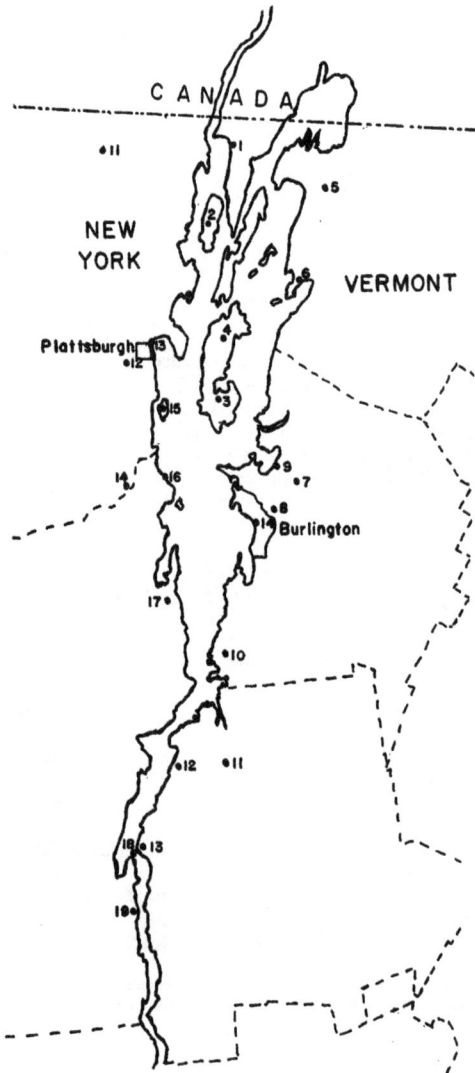

FIG. 29. Sketch map of Lake Champlain area, New York and Vermont, showing marine Pleistocene fossil localities.

*8. Brant Beach, Ocean County. Fill from bottom of Barnegat Bay.

*9. Brigantine Beach, Atlantic County. Hydraulic fill from west side of island.

*10. Somers Point, Atlantic County. Hydraulic fill and shallow wells.

*11. Ocean City, Cape May County. Hydraulic fill on Asbury Avenue between 20th and 21st Streets.

*12. Sea Isle City, Cape May County. Hydraulic fill at "Sea Isle City Gardens."

*13. Peermont, Cape May County. Hydraulic fill.

*14. Avalon, Cape May County. Hydraulic fill.

*15. Wildwood Crest, Cape May County. Hydraulic fill at south end of town.

*16. Two Mile Beach, Cape May County. Hydraulic fill from bottom of former inlet (Turtle Gut Inlet) which separated Wildwood Crest from Two Mile Beach.

*17. Cape May Canal, Cape May County. Excavations from canal dug in 1942. Contains mixture of warm and cold faunas. Probably Cape May formation and Wisconsin. See Richards (1944).

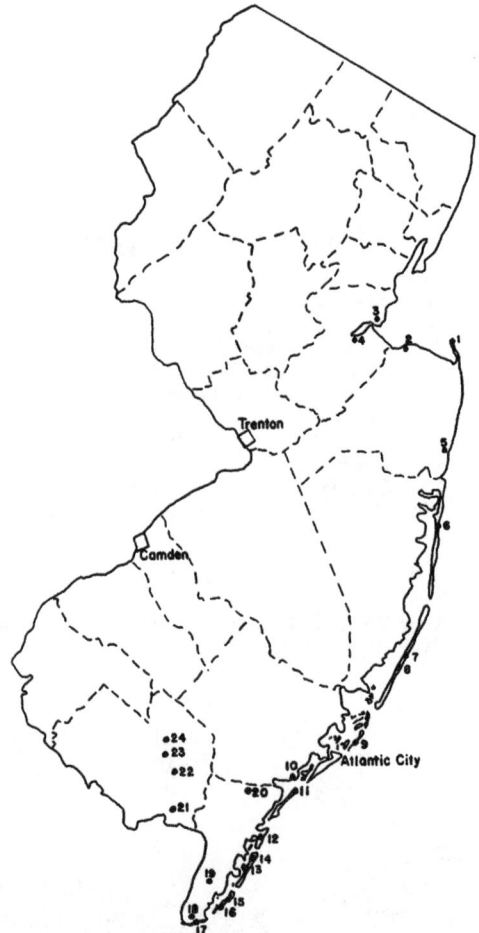

FIG. 30. Sketch map of New Jersey showing marine Pleistocene fossil localities.

*18. Cape May, Cape May County. Hydraulic fill in
Coast Guard Base.
19. Cape May County Airport, near Rio Grande, Cape
May County. Wells between depths of 72 and
193 feet.
20. Tuckahoe, Cape May County. Excavations up to
6 or 8 feet above high water.
21. Heislerville, Cumberland County. Shallow dug
wells.
22. Port Elizabeth, Cumberland County. Shell bed up
to 8 feet above high tide along Maurice River.
23. Buckshutem, Cumberland County. Old records
of shells in excavations on west side of Maurice
River.
24. Millville, Cumberland County, N. J. Excavations
three and one-half miles below Millville; up to 7
feet above high tide.
25. Sediment core from Hudson Canyon, about 225
miles east of mouth of Delaware Bay. Depth of
water 11,400 feet. Probably of Wisconsin age.
Shells carried by turbidity currents (Richards and
Ruhle, 1955).

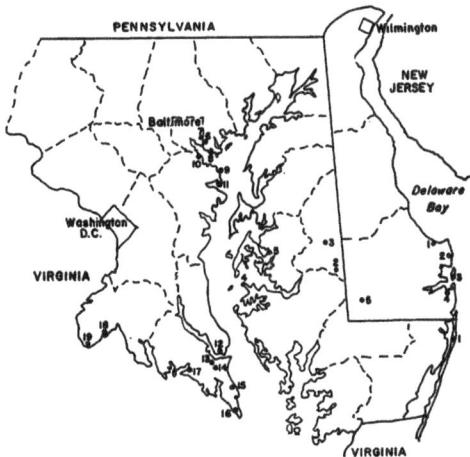

FIG. 31. Sketch map of Delaware and Maryland showing
marine Pleistocene fossil localities.

DELAWARE
(after Richards, 1936)

1. Lewes, Sussex County. Excavations for Lewes-
Rehoboth Canal between Canary and Broadkill
creeks.
*2. Rehoboth, Sussex County. Hydraulic fill adjacent
to Lewes-Rehoboth Canal ("Henlopen Acres").
3. Indian River Inlet, Sussex County. Beach wash.
4. Ocean View, Sussex County. Dredged from
Assawoman Canal.
5. Laurel, Sussex County. Shallow excavations;
elevation +18 feet.

MARYLAND
(after Shattuck, 1906; Smith 1920; Richards, 1936, Blake, 1955)

EASTERN SHORE OF CHESAPEAKE BAY

*1. Ocean City, Worcester County. Hydraulic fill two
miles south of town.
2. Federalsburg, Caroline County. West side of
Marshyhope Creek and at road cut at Herring Hill,
one mile north of town. Shells up to elevation
28 feet.
3. Williston, Caroline County. East bank of Choptank River, one mile north of town. Up to elevation 5 feet.
4. Cooke Point, Dorchester County. Mouth of
Choptank River.
5. Easton, Talbot County. Excavations on Glenwood
Avenue to depth of 8 feet.

WESTERN SHORE OF CHESAPEAKE BAY

6. Back River, Baltimore County. Mouth of Back
River.
7. Middle River, Baltimore County. "Near Middle
River."
8. Sparrows Point, Baltimore County. Well on
north side of Patapsco River, near the mouth.
9. Bodkin Point, Anne Arundel County. South side
of Patapsco River, near mouth. Casts of inde-
terminable Unio.
10. Port Covington, Baltimore County. "Along Patapsco River."
11. Greensbury Point, Anne Arundel County. "Poorly
preserved casts of freshwater Unios . . . on Sev-
ern River opposite Annapolis."
12 Drum Point, Calvert County. North bank of Patuxent River at junction with Chesapeake Bay.
13. Town Creek, St. Mary's County. Mouth of Town
Creek, opposite Solomons Island.
14. South of Town Creek, St. Mary's County. Bank
along Patuxent River.
15. Langleys Bluff, St. Mary's County. Bluff along
Chesapeake Bay between Cedar Point and Point
No Point, about two miles east of Park Hall.
16. Wailes Bluff, St. Mary's County. Bluff on north
bank of Potomac River one mile above Cornfield
Harbor. The bluff rises 15 feet above the beach
and is highly fossiliferous. For detailed descriptions see Richards (1936) and Blake (1955).
17. Near Leonardtown, St. Mary's County. Various
localities between Flood and Poplar Hill Creeks
and along Potomac River a little north of Blake
Creek. See Richards (1936) and Blake (1955).
18. Nanjemoy River, Charles County. East side of
Nanjemoy near mouth.
19. Maryland Point, Charles County. On Potomac
River, just below mouth of the Nanjemoy.

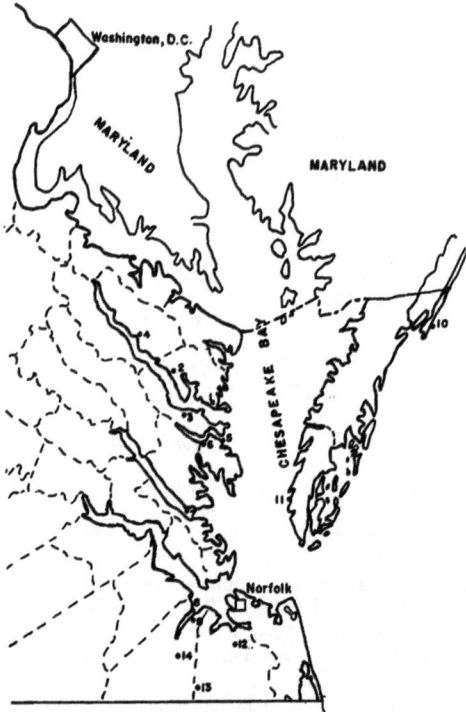

FIG. 32. Sketch map of Virginia showing marine Pleistocene fossil localities.

VIRGINIA

(after Richards, 1936)

1. Near Taft, Lancaster County. North bank of Rappahannock River between Taft and Mosquito Point. 10-foot bluff.
2. Monaskon, Lancaster County. East bank of Rappahannock River, half a mile above Monaskon.
3. Whiting Creek, Middlesex County. Bluffs of Rappahannock River, east of mouth of Whiting Creek.
4. Richardson Creek, Richmond County. Along Rappahannock, near entrance of Richardson Creek.
5. Iron Point, Mathews County. A 15- to 20-foot bluff on Godfrey Bay on left bank of Piankatank River, about half a mile south of Iron Point.
6. Twiggs Ferry, Mathews County. About a quarter of a mile above Green Point Wharf at Twiggs Ferry (or Dixie Post Office) on Piankatank River.
7. Mumford Island, Gloucester County. On northeast bank of York River, opposite Yorktown. Coquina deposit just above tide.
8. Lee's Wharf, Nansemond County. Low bluffs on both sides of Nansemond River at highway bridge eighteen miles below Suffolk. Oyster shells.

9. Gaskins Wharf, Nansemond County. About half a mile above previous locality.
10. Assateague Island, Accomack County. Beach wash.
11. Cape Charles, Northampton County. Hydraulic fill from Chesapeake Bay, half a mile south of town of Cape Charles.

DISMAL SWAMP

12. Near Deep Creek, Norfolk County. Dismal Swamp Canal, five and one-half miles south of Deep Creek.
13. Near Lake Drummond, Norfolk County. Spoil bank along Feeder Canal about a mile east of Lake Drummond.
14. Jericho Canal, Nansemond County. A mixture of Miocene, Pliocene, and Pleistocene species was reported from a road cut near Jericho Canal, two miles east of Suffolk. (The species from this locality are not included in this report.)
15. "Dismal Swamp."

NORTH CAROLINA

(after Richards, 1936, 1950)

1. Intra-Coastal Canal between Pungo and Alligator rivers, Hyde County. Spoil banks at various places along the twenty-two-mile cut. Fossils came from a depth of about 15 feet and are overlain by swamp muck and peat.

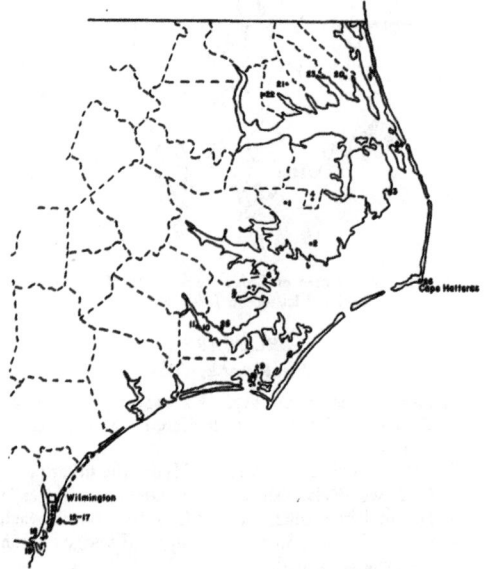

FIG. 33. Sketch map of North Carolina showing marine Pleistocene fossil localities.

2. Lake Matamuskeet, Hyde County. Shallow excavations along shore of lake, especially near Fairfield and New Holland.
3. Stumpy Point, Dare County. Shallow pits along Englehard-Stumpy Point Road, about two miles from Stumpy Point.
4. Kilkenny, Tyrrell County. Shallow excavations on Gum Neck Road, two miles north of Kilkenny.
5. Bayboro, Pamlico County. Excavations north and a little east of Bayboro.
6. Hobucken, Pamlico County. Spoil banks of Intra-Coastal Canal about a mile west of town.
7. Cash Corner, Pamlico County. Excavations from several small canals near Cash Corner and Alliance.
8. Near Beaufort, Cartaret County. Excavations at "Open Land Project" ten miles northwest of Beaufort and six miles from North River; also near Bureau of Fisheries Laboratory at Beaufort.
9. Core Creek Canal, Cartaret County. Excavations for Intra-Coastal Canal between Pamlico Sound and Newport River.
10. Ten miles below New Bern, Craven County. 25-foot bluff along Neuse River near settlement of Croatan.
11. Eleven miles below New Bern, Craven County. Similar bluff. Pleistocene overlies the Pliocene Croatan formation.
12. Gander Point, New Hanover County. Gravel pit between Wilmington and Carolina Beach.
13. Cape Fear River, New Hanover County. Coquina and loose shells were struck in excavations for Intra-Coastal Canal at its junction with the Cape Fear River.
14. Near Carolina Beach, New Hanover County. Excavations for Intra-Coastal Canal at Indian Cove.
15. Old Fort Fisher, New Hanover County. Coquina on beach between tides.
16. Carolina Beach Wharf, New Hanover County. Coquina a mile southeast of Carolina Beach Wharf.
17. Old Brunswick, Brunswick County. Shells in bluff along Cape Fear River, a quarter of a mile north of the ruins of Old Brunswick.
18. Southport, Brunswick County. Well at quarantine station.
19. Fort Caswell, Brunswick County. Well at Fort Caswell, Oak Island, Cape Fear River.
20. South Mills, Camden County. Excavations for Dismal Swamp Canal about five miles north of South Mills north to the Virginia line.
21. Nicanor, Perquimmans County. Excavations for canals near western edge of Dismal Swamp.
22. Bear Swamp, Perquimmans County. Excavations for canals between Hertford and Tyner.
23. Elizabeth City, Pasquotank County. Wells to depth of 30 feet.
24. Manteo, Dare County. Beach wash.

25. Benners Plantation, Pamlico County. East bank of Baird Creek. Left bank of Neuse River, sixteen miles below New Bern.
26. Cape Hatteras, Dare County. Coquina and beach wash.

FIG. 34.   Sketch map of South Carolina showing marine Pleistocene fossil localities.

SOUTH CAROLINA
(after Richards, 1936)

1. Little River, Horry County. Excavations for Intra-Coastal Canal at highway bridge (Route 117) one mile south of Little River.
2. Four miles south of Little River, Horry County. Shells exposed in banks of Intra-Coastal Canal up to 12 feet above low tide (8 feet above high tide). Pleistocene overlies Pliocene Waccamaw formation.
3. Windy Hill, Horry County. Coquina, eleven miles northeast of Myrtle Beach.
4. White Point Creek, Horry County. Shell bed, 6 feet thick, about 5 feet above tide at Price's Creek (= White Point Creek) (after Pugh, 1906).
5. Buckport, Horry County. Oyster shells in ditch one mile southwest of Buckport.
6. Myrtle Beach, Horry County. Coquina on beach.
7. One and one-half miles northwest of Myrtle Beach, Horry County. Banks of Intra-Coastal Canal. Fossiliferous clay exposed up to 6½ feet above low tide overlain by 10 to 13 feet of sand.
8. Laurel, Georgetown County. Bluff along the Waccamaw River (after Pugh, 1906).

9. Waverly Mills, Georgetown County. Excavation to a depth of 19 feet.

10. Winyah Canal, Georgetown County. Excavations for canal connecting Winyah Bay with north run of Santee River.

11. Cooper River, Charleston County. Dredgings making up a small island opposite Etowah Fertilizer Works.

12. Shipyard Creek, Charleston County. Artificial island fill.

13. Stono River, Charleston County. Pleistocene fossils overlying phosphate rock (after Pugh).

14. Charleston, Charleston County.

15. Bees Ferry, Ashley River, Charleston County.

16. Simmons Bluff, Yonges Island, Charleston County. Bluff on Wadmalaw River, one-fourth mile north of Yonges Island railroad station. This locality was studied extensively by Pugh (1906).

17. Beaufort, Beaufort County (after Pugh).

18. Four miles northeast of Monks Corner, Berkeley County. Excavations for Santee-Cooper Diversion Canal. Brackish fauna.

19. Seventeen miles northwest of Monks Corner, Berkeley County. Excavations for Santee-Cooper Diversion Canal Elevation of land 86 feet; elevation of shell bed 65 feet. Mixture of Pliocene and Pleistocene species. Not listed in this report (Richards, 1943).

GEORGIA

(after Richards, 1936)

1. Harners Bridge, Chatham County. Excavations for drainage canal eight miles east of Savannah.

2. Savannah River, Chatham County. Dredgings from Savannah River east of Savannah.

3. Near Savannah, Chatham County.

4. Isle of Hope, Chatham County. 8-foot bluff on Skidway River.

VIII. INTRODUCTION TO THE PALEONTO-LOGICAL SECTIONS

The next two chapters record the pelecypods and gastropods from the Pleistocene deposits between Hudson Bay and Georgia. No attempt has been made to prepare a monograph of the various species, or to cite complete bibliographies. The original reference is given, as well as other significant references, especially those relating to records from the Pleistocene, or those with suitable illustrations. The distribution of the various species is given, with locality numbers as itemized in chapter VII. In a few cases where the exact locality is not known, merely the state is listed. Records from outside the limits of this report, for example Florida, are listed merely by state. The data on the Pleistocene are followed by information on the present range of the species, taken largely from the works of Johnson (1934), LaRoque (1943) and Abbott (1954), as well as from information obtained from the collections of the Academy of Natural Sciences.

To save space, references that are cited many times are not given in full, but merely by date, the full reference being given in the bibliography. These include the following works: Blake (1953), Clark (1906), Dall 1898–1903), Feyling-Hanssen (1955), Gardner (1943), Holmes (1860), and Olsson and Harbison (1953).

Almost all the species are illustrated. When possible, Pleistocene specimens have been figured. However, when the records are based upon damaged or poorly preserved individuals, photographs of Recent specimens have been used. This was done in order to render the pictures more useful for identification.

No pretense is made of including every species known from the Pleistocene of the area in question. Some

Fig. 35. Sketch map of Georgia showing marine Pleistocene fossil localities.

records which have not been confirmed have been omitted. All representatives of certain genera of rather small gastropods have not been considered. These include *Lora* (= *Bela*), *Turbonilla,* and *Odostomia.* The identification of many of these forms is exceedingly difficult unless perfect material is available, and many of the species are not well defined. Furthermore, they are of little value in stratigraphic studies.

In addition to material obtained by the writer in the field, collections were examined at the Academy of Natural Sciences, the U. S. National Museum, the Museum of Comparative Zoology, (Cambridge) the Redpath Museum (Montreal), the National Museum of Canada (Ottawa), the American Museum of Natural History (New York), Vanderbilt University (Nashville, Tenn.), as well as various museums in London, Copenhagen, Oslo, Stockholm, Helsinki, Moscow, and Leningrad.

IX. PELECYPODA

Family NUCULIDAE

Nucula expansa Reeve, 1855

Plate 1, Figures 2–5

Nucula expansa Reeve, 1855, *Last of the Arctic Voyages,* app., 397.

Pleistocene Distribution:
Newfoundland: 3, 8, 15
Quebec: 4
New Brunswick: 3
Maine: 3, 19, 20, 21
Present Distribution: Labrador and north shore Gulf of St. Lawrence; 30 fathoms.

Nucula tenuis (Montagu), 1808

Plate 1, Figures 6–9

Arca tenuis Montagu, 1808, *Test. Brit. Suppl.,* 56, pl. 29, fig. 1.

Pleistocene Distribution:
Quebec: 4, 9
Maine: 3
Present Distribution: Labrador to North Carolina; 4 to 100 fathoms.

According to Ockelmann (1958) *N. expansa* is a variety of *N. tenuis.*

Nucula proxima Say, 1820

Plate 1, Figure 1

Nucula proxima Say, 1820, *Amer. Jour. Sci.,* 1st ser. 2: 40.
Nucula proxima Holmes, 1860: 17, pl. 3, fig. 6.
Nucula proxima Clark, 1906: 207, pl. 65, figs. 1–4.
Nucula proxima Gardner, 1943: 19, pl. 1, figs. 1, 2, 4, 5.
Nucula proxima Olsson, 1953: 27.

Pleistocene Distribution:
New York: 2, 3
New Jersey: 21, 25
Maryland: 15, 16
Virginia: 7, 15
North Carolina: 1, 9, 10, 11, 12, 18
South Carolina: 2, 12, 16
Florida
Present Distribution: Nova Scotia to Texas.

Nucula major Richards, 1944

Plate 1, Figures 10, 11

Nucula major Richards, 1944, *Acad. Nat. Sci. Phila., Notula Naturae,* no. 134: 8, figs. 5, 6.
Nucula major Richards, 1947, *Jour. Paleont.* 21: 31, pl. 12, figs. 1, 2.

Pleistocene Distribution:
New Jersey: 17
Present Distribution: Extinct. Closely related to *N. shaleri* Dall from the Miocene of Martha's Vineyard, Massachusetts.

Nuculana buccata (Steenstrup), 1842

Plate 1, Figures 21, 22

Leda buccata Steenstrup, 1842, *Index Moll. Groenl.,* 17.

Pleistocene Distribution:
Hudson Bay: 1, 3, 6
New Brunswick: 1, 2
Maine: 2, 4
Present Distribution: Greenland.

Nuculana tenuisulcata (Couthouy), 1838

Plate 1, Figure 17

Nucula tenuisulcata Couthouy, 1838, *Bost. Jour. Nat. Hist.* 2: 64.

Pleistocene Distribution:
Maine: 2, 3, 9
Present Distribution: Arctic to Rhode Island.

Nuculana pernula (Müller), 1779

Plate 1, Figures 14, 15

Arca pernula Müller, 1779, *Beschäft Naturf. Freunde zu Berlin* 4: 55.

Pleistocene Distribution:
Labrador: 2, 5
Hudson Bay: 1
James Bay: 2, 3, 4, 5, 6, 7
Newfoundland: 2, 3, 6, 7, 11, 25
New Brunswick: 1
Quebec: 4
Maine: 2, 3
Present Distribution: Arctic to Cape Cod and northern Alaska.

Nuculana jacksoni (Gould), 1941

Plate 1, Figures 29, 30

Nucula jacksoni Gould, 1841, *Rept. Invert. Mass.*, 102.

Pleistocene Distribution:
  Maine: 2, 20
Present Distribution: Labrador; 10–80 fathoms.

This species is closely applied to *N. pernula* and may be only a variety.

Nuculana minuta (Müller), 1776

Plate 1, Figure 16

Arca minuta Müller, 1776, *Zoologica danicae Prodr.*, 247.
Arca minuta Fabricius, 1780, *Fauna Groenl.*, 414.

Pleistocene Distribution:
  Labrador: 1, 2, 5
  Quebec: 4, 9
Present Distribution: Arctic to Maine.

Nuculana acuta (Conrad), 1832

Plate 1, Figures 12, 13

Nucula acuta Conrad, 1832, *Amer. Marine Conch.*, 32, pl. 5, fig. 1; pl. 6, fig. 3.
Nucula acuta Holmes, 1860: 16, pl. 3, fig. 7.
Leda acuta Clark, 1906: 208, pl. 65, figs. 5–8.
Nuculana acuta Olsson, 1953: 28.

Pleistocene Distribution:
  Maryland: 15, 16
  Virginia: 16
  North Carolina: 9, 10, 11, 18
  South Carolina: 14
  Florida, Alabama, Mississippi, Louisiana, Texas
Present Distribution: Massachusetts to West Indies and Texas.

Yoldia (Portlandia) glacialis (Wood), 1828

Plate 1, Figures 18–20, 27, 28

Nucula arctica Gray, 1824, *Captain Parry's First Voyage*, suppl. to app., 241.[3]
Nucula glacialis Wood, 1828, *Suppl. Index Test.*, 45.

Pleistocene Distribution:
  Newfoundland: 13, 26
  Quebec: 4, 5, 7, 9
  Ontario: 4
  New Brunswick: 1, 2, 3
  Maine: 1, 2, 3, 4, 8, 12, 13, 15, 16, 17, 19, 20, 21
  Vermont: 6, 9, 13
Present Distribution: Arctic.

---

[3] This name, used by many authors, is based upon Gray's inadequate description of an unsculptured clam with a taxodont hinge.

Yoldia (Yoldiella) lenticula (Möller), 1842

Plate 1, Figures 23, 24

Nucula lenticula Möller, 1842, *Index Moll. Groenl.*, 17.
Nucula pygmaea Philippi, 1842. Not von Munster, 1835.
Yoldia abyssicola Torrell, 1859. Not Sars, 1859.

Pleistocene Distribution:
  James Bay: 1
  Ontario: 3
  Maine: 2, 3
  New Jersey: 25
Present Distribution: North of Cape Cod; deep water.

Yoldia limatula (Say), 1831

Plate 1, Figure 25

Nucula limatula Say, 1831, *Amer. Conch.*, pl. 12.
Leda limatula Holmes, 1860: 18, pl. 3, fig. 8.
Yoldia limatula Clark, 1906: 209, pl. 65, figs. 9–12.
Yoldia limatula Blake, 1953: 24.

Pleistocene Distribution:
  Quebec: 4
  New York: 8
  New Jersey: 25
  Maryland: 16
  North Carolina: 1, 3, 10, 18
  South Carolina: 16
Present Distribution: Gulf of St. Lawrence to North Carolina.

Yoldia myalis Couthouy, 1838

Plate 1, Figure 26

Yoldia myalis Couthouy, 1838, *Boston Jour. Nat. Hist.* **2**: 62.

Pleistocene Distribution:
  Labrador: 5 (after Packard)
  Maine: 3
Present Distribution: Labrador to Cape Cod; Alaska.

Family ARCIDAE

Arca (Bathyarca) glacialis Gray, 1824

Plate 2, Figures 17, 18

Arca glacialis Gray, 1824, *Parry's First Voyage, 1819–1820*, supp. and app., 244.
Arca glacialis Sheldon, 1916, *Paleont. Amer.* **1**: 65, pl. 16, figs. 12–14.

Pleistocene Distribution:
  Maine
Present Distribution: Gulf of Saint Lawrence, Greenland.

Arca zebra Swainson, 1833

Plate 3, Figures 1, 3

Arca zebra Swainson, 1833, *Zool. Illust. Shells*, 2nd. ser., 3(26): pl. 118.
Arca occidentalis Philippi, 1847, *Abbild. Conch.* **3**: 29, pl. 4, figs. 4a–4c.

Arca occidentalis Sheldon, 1916, *Paleont. Amer.* 1: 8, pl. 1, figs. 8–11.
Arca zebra Abbott, 1958, *Acad. Nat. Sci. Monograph* 11: 109.

Pleistocene Distribution:
　North Carolina: 26
　South Carolina: 4
Present Distribution: North Carolina to West Indies.

Arca adamsi Smith, 1890

Plate 2, Figures 14–16

Arca coelata Conrad, 1845, *Foss. Medial Tertiary*, 61, pl. 32, fig. 2. Not *Arca coelata* Reeve, 1844, *Conch. Icon., Arca* no. 110.
Arca adamsi Smith, 1890, *Jour. Linn. Soc. Zool.* 20: 499, pl. 30, figs. 6, 6a.
Arca caelata Holmes, 1860: 22, pl. 4, figs. 6, 6a.
Arca adamsi (Shuttleworth manuscript) Smith, Sheldon, 1916, *Paleont. Amer.* 1(1): 22, pl. 4, figs. 16–18; pl. 5, fig. 1.

Pleistocene Distribution:
　South Carolina: 14
Present Distribution: North Carolina to West Indies.

Anadara notabilis (Röding), 1798

Plate 1, Figures 31, 32

Arca notabilis Röding, 1798, *Museum Boltonianum*, 173.
Arca auriculata of authors, not Lamarck, 1819 (which is from the Red Sea).
Arca deshayesi Hanley, 1843, *Catologue Recent Bivalve Shells*, 157.
Arca notabilis Abbott, 1958, *Acad. Nat. Sci. Phila., Monograph* 11: 111.

Pleistocene Distribution:
　North Carolina: 26
Present Distribution: Northern Florida to Brazil.

Anadara lienosa (Say), 1832

Plate 2, Figures 7, 8

Arca lienosa Say, 1832, *Amer. Conch.*, pl. 36, fig. 1.
Arca secticostata Reeve, 1844, *Conch. Inconica* II *Arca* No. 38.
Arca lienosa Say, Holmes, 1860: 20, pl. 4, figs. 3, 3a.
Arca lienosa Sheldon, 1916, *Paleont. Amer.* 1: 35, pl. 7, figs. 26–28; pl. 8, figs. 1, 2.
Anadara lienosa Gardner, 1943: 23, pl. 2, figs. 4, 7.

Pleistocene Distribution:
　North Carolina: 26 ?
　South Carolina: 16
Present Distribution: North Carolina to Greater Antilles.

Anadara transversa (Say), 1822

Plate 2, Figures 12–13

Arca transversa Say, 1822, *Jour. Acad. Nat. Sci. Phila.* 2: 296.
Arca transversa Holmes, 1860: 21, pl. 4, figs. 5–5a.
Arca transversa Clark, 1906: 206, pl. 64, figs. 7–10.

Pleistocene Distribution:
　Massachusetts: 1
　New York: 2, 3, 6
　New Jersey: 3, 11, 12, 13, 16, 17
　Maryland: 2, 14, 15, 16
　North Carolina: 1, 3, 5, 6, 7, 8, 9, 10, 11, 13, 21, 26
Present Distribution: Massachusetts to Texas.

Anadara brasiliana Lamarck, 1819

Plate 2, Figures 5, 6

Arca brasiliana Lamarck, 1819, *Animaux sans Vert.* 6: 44.
Arca incongrua Say, 1822, *Jour. Acad. Nat. Sci. Phila.* 2: 268.
Arca incongrua Holmes, 1860: 19, pl. 4, figs. 1, 1a.
Arca incrongrua Say, Sheldon, 1916, *Paleont. Amer.* 1: 59, pl. 14, figs. 4–7.

Pleistocene Distribution:
　North Carolina: 3, 10, 12, 13, 15, 26
　South Carolina: 4, 11, 13, 14, 16
　Georgia: 1, 2
Present Distribution: North Carolina to West Indies and Texas.

Anadara ovalis Bruguiere, 1789

Plate 2, Figures 3, 4

Arca ovalis Bruguiere, 1789, *Ency. Méth.*, 110.
Arca campechiensis Gmelin, 1790, *Syst. Nat.*, 13th ed., 1: 3312.
Arca americana Reeve, 1844, *Conch. Icon., Arca*, fig. 21.
Arca americana Gray, Holmes, 1860: 19, pl. 4, figs. 2–2a.

Pleistocene Distribution:
　Massachusetts: 1
　New York: 2, 3
　New Jersey: 5, 8, 9, 11, 12, 13, 16, 17
　Delaware: 1, 3
　Maryland: 1
　Virginia: 7, 12, 15
　North Carolina: 1, 5, 7, 8, 9, 10, 13, 16, 26
　South Carolina: 2, 7, 8, 11, 12, 13, 14, 16
Present Distribution: Massachusetts to West Indies and Gulf states.

Eontia palmerae MacNeil, 1938

Plate 2, Figures 10, 11

Eontia palmerae MacNeil, 1938, *U. S. Geol. Surv. Prof. Paper* **189A**: 23, pl. 3, figs. 7, 8.

Pleistocene Distribution:
　New Jersey: Heislerville
　Maryland: Wailes Bluff (type locality); Potomac River, St. Mary's Co.
　North Carolina: Dismal Swamp Canal, 3.5 miles south of Virginia line
Present Distribution: Extinct, close to *E. ponderosa* (Say).

The Heislerville specimens were regarded as Pliocene by Dall, and as "probably not very late Pleistocene" by MacNeil.

Eontia veroensis MacNeil, 1938

Plate 2, Figure 9

Eontia veroensis MacNeil, 1938, *U. S. Geol. Surv. Prof. Paper*
    189A: 24, pl. 3, fig. 6.

Pleistocene Distribution:
    Florida: Vero, St. Lucie County (type locality)
Present Distribution: This is also an extinct species
and is regarded as "not very late Pleistocene" by
MacNeil.

Noetia (Eontia) ponderosa (Say), 1822

Plate 2, Figures 1, 2

Arca ponderosa Say, 1822, *Jour. Acad. Nat. Sci. Phila.* 2: 267.
Arca ponderosa Holmes, 1860: 21, pl. 4, figs. 4–4a.
Arca ponderosa Clark, 1906: 205, pl. 64, figs. 1–6.
Arca ponderosa Sheldon, 1916, *Paleont. Amer.* 1: 28, pl. 6, figs.
    6–10 (in part).
Eontia ponderosa MacNeil, 1938, *U. S. Geol. Surv. Prof. Paper*
    189A: 24, pl. 3, figs. 9–12.
Arca ponderosa Blake, 1953: 24.

Pleistocene Distribution:
    Massachusetts: 1
    New Jersey: 9, 16, 17
    Maryland: 1, 15, 16
    Virginia: 7, 10, 11, 15
    North Carolina: 1, 3, 6, 7, 8, 9, 10, 12, 13, 15, 16, 26
    South Carolina: 1, 2, 8, 11, 14, 16
    Georgia: 1, 2
Present Distribution: Virginia to Gulf of Mexico.

The affinities of *E. ponderosa* (Say) and the closely
related Pleistocene *E. palmerae* MacNeil as well as the
Pliocene *E. variabilis* MacNeil, *E. tillensis* MacNeil and
E. limula (Conrad) are fully discussed by MacNeil
(1938).

Glycymeris americana (DeFrance), 1826

Plate 3, Figures 8, 9

Pectunculus americanus DeFrance, 1826, *Dict. Science Nat.* 39:
    225.
? *Pectunculus carolinensis* Holmes, 1860: 15, pl. 3, fig. 4.
Glycymeris americana Gardner, 1943: 27, pl. 1, figs. 16–21.

Pleistocene Distribution:
    North Carolina: 26
    South Carolina: 16
    Florida
Present Distribution: North Carolina to the West
Indies and Texas.

Glycymeris charlestonensis (Holmes), 1860

Pectunculus charlestonensis Holmes, 1860: 16, pl. 3, fig. 5.

Pleistocene Distribution:
    South Carolina: (after Holmes)
Present Distribution: Known only from Holmes's
record.

Glycymeris pectinata (Gmelin), 1790

Plate 3, Figures 6, 7

Arca pectinata Gmelin, 1790, *Syst. Nat.*, 13th ed., 3313.

Pleistocene Distribution:
    South Carolina: 13, 14, 16
    Florida, Louisiana
Present Distribution: North Carolina to both sides of
Florida and the West Indies.

FAMILY PTERIIDAE

Pteria colymbus (Röding), 1798

Plate 3, Figure 5

Pinctata colymbus Röding, 1798, *Museum Boltenianum* 2: 106.
Avicula atlantica Holmes, 1860: 14.
Pteria colymbus Dall, 1898: 670.
Pteria colymbus Olsson, 1953: 44.

Pleistocene Distribution:
    South Carolina: 14, 16
Present Distribution: North Carolina to southeast
Florida and West Indies.

FAMILY PINNIDAE

Atrina rigida (Solander), 1786

Plate 3, Figure 2

Pinna rigida Solander, 1786, *Catalogue of the Portland Museum*,
    136, species 3040.
Pinna rigida Dillwyn, 1817, *Catalogue*, 327.
Pinna carolinensis Hanley, 1858, *Proc. Zool. Soc. London*, 225.
Pinna seminuda Holmes, 1860, *Post Pleistocene South Carolina*,
    14, pl. 3, fig. 2. Not Lamarck.
Atrina rigida Dall, 1895: 663.
Atrina (Atrina) rigida Turner and Rosewater, 1958, *Johnsonia*
    3(38): 312, pl. 158, 159, figs. 1–4.

Pleistocene Distribution:
    South Carolina: 14, 16
    Florida
Present Distribution: North Carolina to south half
of Florida and the Caribbean.

This species has been frequently confused with *A.
seminuda* (Lamarck).

Atrina serrata (Sowerby), 1825

Plate 3, Figure 4

Pinna serrata Solander, 1786, *Catalogue of the Portland Mu-
    seum*, 71, 165 (*nomen nudum*).
Pinna serrata Sowerby, 1825, *Catalogue of Shells of Earl of
    Tankerville, London*, 23, app., p. v.
Pinna seminuda Lamarck, Reeve, 1841, *Conch. Iconica* 11, *Pinna*,
    pl. 2, fig. 2. Not *Pinna seminuda* Lamarck, 1819.
Pinna muricata Linné, Holmes, 1860, *Post Pleiocene Fossils of
    South Carolina*, 15, pl. 3, fig. 3. Not *Pinna muricata* Linné,
    1758.
Atrina (Servatrina) serrata Turner and Rosewater, 1958,
    Johnsonia 3(38): 320, pls. 170, 171.

Pleistocene Distribution:
South Carolina: 16
Louisiana
Present Distribution: North Carolina and south half of Florida.

FAMILY PECTINIDAE

Aquipecten irradians (Lamarck), 1819

Plate 3, Figure 10; Plate 4, Figure 3

Ostrea gibba Linné, 1758, *Syst. Nat.,* 10th ed., 698 (part).
Pecten irradians Lamarck, 1819, *Animaux sans Vert.,* 1st ed., 6: 173.
Pecten dislocatus Say, 1822, *Jour. Acad. Nat. Sci. Phila.,* ser. 1, 2: 260.
Pecten dislocatus Holmes, 1860: 12, pl. 2, fig. 12.
Pecten gibbus Gardner, 1943: 31, pl. 5, fig. 3.

Pleistocene Distribution:
New York: 2, 7
New Jersey: 5, 6, 7, 9, 11, 17
Delaware: 3
Maryland: 1
North Carolina: 1, 4, 6, 8
South Carolina: 16
Present Distribution: Nova Scotia to Texas and West Indies.

This is a variable form, sometimes divided into several varieties or distinct species. The name *P. irradians* is in general use for the form common from North Carolina northward. *Pecten gibbus, P. nucleus,* and *P. dislocatus* are slightly more rounded and are known from more southerly waters.

Aequipecten muscosus (Wood), 1828

Plate 4, Figure 2

Ostrea muscosus Wood, 1828, *Suppl. Index Testacea,* 47.
Pecten exasperatus Sowerby, 1842, *Thes. Conch.* 1: 54, pl. 18, figs. 183–186.

Pleistocene Distribution:
South Carolina: ? (after Pugh)
Present Distribution: North Carolina to Florida, Texas, and West Indies.

Placopecten magellanicus (Gmelin), 1791

Plate 4, Figure 4; Plate 9, Figure 19

Pecten grandis Solander, 1786, *Portland Catalogue (nomen nudum).*
Ostrea magellanicus Gmelin, 1791, *Syst. Nat.,* 13th ed., 3317.

Pleistocene Distribution:
New Brunswick: 3
New York: 3, 5
New Jersey: 25
Present Distribution: Labrador to North Carolina; 10 to 100 fathoms.

Pecten groenlandica Sowerby, 1842

Plate 3, Figures 15, 16

Pecten groenlandica Sowerby, 1842, *Thes. Conch.* 1(2): 57.

Pleistocene Distribution:
Maine: 2, 3
Present Distribution: Greenland to Newfoundland.

Closely related to *P. magellanicus* (Gmelin), and may represent merely young individuals of that species.

Chlamys islandicus (Müller), 1776

Plate 4, Figure 1

Pecten islandicus Müller, 1776, *Zool. Can. Prodr.* no. 2990: 248.
Chlamys islandica Feyling-Hanssen, 1955: 128, pl. 18, figs. 1–3.

Pleistocene Distribution:
Hudson Bay: 1, 3, 4, 6
James Bay: 2, 3
Labrador: 1
Newfoundland: 3, 5, 6, 8, 16, 17, 22
Quebec: 4, 7
New Brunswick: 1, 3
Maine: 2, 4, 6, 7, 15, 19
New York: 3
Present Distribution: Arctic to Massachusetts. Alaska to Washington.

FAMILY SPONDYLIDAE

Plicatula gibbosa Lamarck, 1801

Plate 3, Figures 11, 12

Plicatula gibbosa Lamarck, 1801, *Animaux sans Vert.,* 132.
Plicatula cristata Holmes, 1860: 13, pl. 2, fig. 13.

Pleistocene Distribution:
South Carolina: 16
Present Distribution: North Carolina to Gulf states and West Indies.

FAMILY OSTREIDAE

Crassostrea virginica (Gmelin), 1791

Plate 4, Figures 5–7

Ostrea virginica Gmelin, 1791, *Syst. Nat.,* 13th ed. 1(6): 3336.
Ostrea virginiana var. *procyon* Holmes, 1860: 10, pl. 2, fig. 9a.
Ostrea virginiana Holmes, 1860: 9, pl. 2, fig. 9.
Ostrea virginiana Dall, 1898: 687.
Ostrea virginiana Clark, 1906: 204, pl. 61, 62, 63.

Pleistocene Distribution:
Massachusetts: 1, 3
New York: 2, 3, 6
New Jersey: 2, 5, 7, 8, 9, 10, 11, 12, 13, 16, 17, 18, 19, 20, 21, 22, 23
Delaware: 1, 2, 3, 4, 5
Maryland: 1, 2, 3, 4, 12, 13, 15, 16, 17, 19
Virginia: 1, 2, 3, 4, 12, 13, 15, 16, 17, 19

North Carolina: 1, 5, 6, 7, 8, 9, 10, 12, 13, 15, 21, 22
South Carolina: 1, 2, 3, 4, 5, 12, 14, 16, 18
Georgia: 1, 2, 3, 4
Florida, Alabama, Louisiana, Texas
Present Distribution: Gulf of St. Lawrence to Gulf of Mexico and West Indies.

FAMILY **LIMIDAE**

Lima scabra (Born), 1778

Plate 4, Figures 9, 10

Ostrea scabra Born, 1778, *Test. Mus. Vindobonensis,* 110.
Lima (Ctenoides) scabra Dall, 1898, pt. 4: 768.
Lima (Ctenoides) scabra Olsson, 1953: 59.

Pleistocene Distribution:
South Carolina: ? (after Pugh)
Present Distribution: Southern Florida to the West Indies.

FAMILY **ANOMIIDAE**

Anomia simplex Orbigny, 1845

Plate 4, Figure 18; Plate 5, Figure 22

Anomia simplex Orbigny, 1845, *Moll. Cubana,* 367, pl. 38, figs. 31–33.
Anomia ephippium Linné, Holmes, 1860: 11, pl. 2, fig. 11.

Pleistocene Distribution:
Massachusetts: 1
New York: 2, 3
New Jersey: 3, 5, 8, 9, 11, 12, 13, 16, 17
Delaware: 1, 3
Maryland: 1
North Carolina: 1, 8, 10, 11
South Carolina: 2, 11, 14, 16
Present Distribution: Nova Scotia to West Indies.

Anomia aculeata Müller, 1776

Plate 4, Figures 15, 16

Anomia aculeata Müller, 1776, *Z. Dan. Prod.,* 249.

Pleistocene Distribution:
Massachusetts: 1
Present Distribution: Labrador to North Carolina.

FAMILY **MYTILIDAE**

Modiolus modiolus (Linné), 1758

Plate 4, Figure 8

Mytilus modiolus Linné, 1758, *Syst. Nat.,* 10th ed., 706.
Volsella modiolus Blake, 1953: 24.
Volsella modiola Feyling-Hanssen, 1955: 133, pl. 19, figs. 1–3.

Pleistocene Distribution:
Newfoundland: 3
Quebec: 9
Massachusetts: 1, 3

New Jersey: 17
Maryland: 16
South Carolina: 14, 16
Present Distribution: Arctic to northeast Florida. Arctic to San Pedro, California.

Modiolus demissus (Dillwyn), 1817

Plate 4, Figure 19

Mytilus demissus Dillwyn, 1817, *Catalogue* 1: 314.

Pleistocene Distribution:
New Jersey: 2, 7, 9, 11, 13, 16, 17
Delaware: 1, 2
Maryland: 1
Virginia: 9
South Carolina: 2
Present Distribution: Gulf of St. Lawrence to South Carolina. Introduced to California.

Mytilus edulis Linné, 1758

Plate 4, Figures 11, 12

Mytilus edulis Linné, 1758, *Syst. Nat.,* 10th ed., 705.
Mytilus edulis Feyling-Hanssen, 1955: 130, pl. 18, figs. 4, 5.

Pleistocene Distribution:
Labrador: 1
Hudson Bay: 1, 2, 5
Newfoundland: 5, 6, 8, 16, 17, 18, 24, 38
Quebec: 4, 7, 9, 13
New Brunswick: 1, 3
Maine: 2, 13, 19, 20
Vermont: 2, 3, 5, 6, 9
New York: 2
New Jersey: 2, 3, 5, 16, 17
Delaware: 3
South Carolina: ? (after Pugh)
Present Distribution: Arctic to South Carolina.

Brachidontes exustus (Linné), 1758

Plate 4, Figures 13, 14

Mytilus exustus Linné, 1758, *Syst. Nat.,* 10th ed., pl. 705, no. 213.
Brachidontes exustus Olsson, 1953: 62.

Pleistocene Distribution:
South Carolina: 16
Florida
Present Distribution: North Carolina to Texas and West Indies.

Brachidontes recurvus (Rafinesque), 1820

Plate 4, Figure 17

Mytilus recurvus Rafinesque, 1820, *Annales gen. sci. phys. Bruxelles* 5: 320.
Mytilus hamatus Say, 1822, *Jour. Acad. Nat. Sci. Phila.* 2: 265.
Mytilus hamatus Clark, 1906: 203, pl. 60, figs. 5, 6.
Brachidontes recurvus Gardner, 1943: 29, pl. 1, figs. 7, 8.
Mytilus recurvus Blake, 1953: 24.

Pleistocene Distribution:
Massachusetts: 1
New Jersey: 18
Maryland: 2, 16
Virginia: 9
North Carolina: 5
Present Distribution: Cape Cod to Texas and West Indies.

Crenella glandula Totten, 1834

Plate 5, Figures 3, 4

Crenella glandula Totten, 1834, *Amer. Jour. Sci.* **26**: 367.

Pleistocene Distribution:
Quebec: 9 ? (Dawson)
Massachusetts: 1
New Jersey: 17
Present Distribution: Labrador to North Carolina.

Musculus corrugatus (Stimpson), 1851

Plate 5, Figures 5, 6

Mytilus corrugata Stimpson, 1851, *Shells of New England*, 12 (no desc.).

Pleistocene Distribution:
Quebec: 4
Maine: 12
Present Distribution: Greenland to North Carolina.

Musculus lateralis (Say), 1822

Plate 5, Figures 7, 8

Mytilus lateralis Say, 1822, *Jour. Acad. Nat. Sci. Phila.* **2**: 264.

Pleistocene Distribution:
South Carolina: 16
Present Distribution: Delaware to West Indies.

Musculus substriata (Gray), 1824

Plate 5, Figures 9, 10

Modiolaria substriata Gray, 1824, *Parry's First Arct. Voyage*, app., 245.
Modiolaria discors Gould, 1841. Not Linné.
Modiolaria substriata LaRoque, 1953: 39.
Musculus discors Abbott, 1954, *American Seashells*, 355.
Musculus discors substriatus Feyling-Hanssen, 1955: 134, pl. 19, figs. 4–7.

Pleistocene Distribution:
Quebec: 6
Maine: 19, 20
Present Distribution: Arctic to Long Island Sound. Greenland.

Musculus nigra (Gray), 1824

Plate 5, Figure 12

Modiola nigra Gray, 1824, *Parry's First Voyage*, suppl. to app. 244.
Modiolaria nigra of authors.

Pleistocene Distribution:
Quebec: 4, 9
Labrador: 1
Maine: 13
Present Distribution: Circumpolar to North Carolina.

FAMILY DREISSENIIDAE

Congeria leucopheatus (Conrad), 1831

Plate 5, Figures 1, 2

Mytilus leucopheatus Conrad, 1831, *Jour. Acad. Nat. Sci. Phila.* **6**: 263, pl. 11, fig. 13.

Pleistocene Distribution:
North Carolina: 1
Present Distribution: New York to Mexico.

FAMILY THRACIIDAE

Thracia septentrionalis Jeffries, 1872

Plate 5, Figure 11

Thracia septentrionalis Jeffreys, 1872, *Ann. Mag. Nat. Hist.*, ser. 4, **10**: 238.

Pleistocene Distribution:
Maine: 15
Massachusetts: 1
Present Distribution: Greenland to off Block Island.

Thracia conradi Couthouy, 1839

Plate 5, Figures 13, 14

Thracia declivis Conrad, 1831, *Amer. Marine Conch.* **2**: 44, pl. 9, fig. 2. Not *Mya declivis* Pennant, 1778.
Thracia conradi Couthouy, 1839, *Boston Jour. Nat. Hist.* **2**: 153, pl. 4, fig. 2.
Thracia conradi Dall, 1903: 1524.
Thracia conradi Gardner, 1943: 43, pl. 10, fig. 4.

Pleistocene Distribution:
Maine: 3
New York: 3
Present Distribution: Nova Scotia to Long Island Sound.

FAMILY LYONSIIDAE

Lyonsia arenosa (Möller), 1842

Plate 5, Figure 17

Pandorina arenosa Möller, 1842, *Index Moll. Groenl.*, 20.

Pleistocene Distribution:
Quebec: 2, 3, 4, 9
New Brunswick: 2
Maine: 3
Present Distribution: Alaska to Vancouver; Greenland to Maine, Cape Ann.

Lyonsia hyalina (Conrad), 1831

Plate 5, Figure 15

Mya hyalina Conrad, 1831, *Jour. Acad. Nat. Sci. Phila.* **6**: 261, pl. 11, fig. 12.
Lyonsia hyalina Blake, 1953: 25.

Pleistocene Distribution:
  Maryland: 15
Present Distribution: Gulf of St. Lawrence to South Carolina, Texas.

Family PANDORIDAE

Pandora gouldiana Dall, 1886

Plate 5, Figure 18; Plate 6, Figures 3, 4

Pandora trilineata Conrad, 1831, *Amer. Marine Conch.,* 49, pl. 10, figs. 1, 2. Not of Say, 1822.
Pandora gouldiana Dall, 1886, *Bull. Mus. Comp. Zool.* **12**: 312.
Pandora gouldiana Dall, 1903: 1521.
Pandora gouldiana Blake: 1953: 25.

Pleistocene Distribution:
  Massachusetts: 1
  New Jersey: 17
  Maryland: 16
Present Distribution: Gulf of St. Lawrence to New Jersey.

Pandora arctica Dall, 1903

Plate 5, Figure 19

Pandora arctica Dall, 1903, *Trans. Wagner Free Inst. Sci.* **3**(6): 1520, pl. 57, fig. 26.

Pleistocene Distribution:
  Maine: 3
Present Distribution: Arctic.

Pandora glacialis Leach, 1819

Plate 6, Figures 1, 2

Pandora glacialis Leach, 1819, J. Ross, *Voyage Discovery Baffin's Bay,* ed. 2, app. 4, p. 174.
Pandora glacialis Feyling-Hanssen, 1955: 151, pl. 19, figs. 8, 9.

Pleistocene Distribution:
  James Bay: 5
  Quebec: 9
  Maine: 3, 9
Present Distribution: Circumpolar, Arctic to Massachusetts; 45 to 100 fathoms.

Pandora trilineata Say, 1822

Plate 6, Figure 5

Pandora trilineata Say, 1822, *Jour. Acad. Nat. Sci. Phila.,* 1st ser., **2**: 261.
Pandora trilineata Dall, 1903: 1519.
Pandora trilineata Gardner, 1943: 49, pl. 11, fig. 7.

Pleistocene Distribution:
  New Jersey: 17
  Virginia: 15
  North Carolina: 10, 11
  South Carolina: 16
Present Distribution: North Carolina to Texas.

Pandora arenosa Conrad, 1834

Plate 5, Figures 20, 21

Pandora arenosa Conrad, 1834, *Jour. Acad. Nat. Sci. Phila.,* 1st ser., **7**: 130.
Pandora arenosa Dall, 1903: 1518.
Pandora arenosa Gardner, 1943: 45, pl. 10, figs. 16, 19, 20.

Pleistocene Distribution:
  Maine: (?) Packard
  North Carolina: 1
Present Distribution: North Carolina to southeastern Florida; 7 to 48 fathoms.

Family ASTARTIDAE

Astarte banksii (Leach), 1819

Plate 6, Figures 14, 15

Venus compressa Montagu, 1808, *Testacea Britannica,* suppl., 43, pl. 26, fig. 1 (in part). Not Linné.
Venus montagui Dillwyn, 1817, *Descriptive Catalogue of Recent Shells,* 167 (in part).
Nicania banksii Leach, 1819, in J. Ross, *Voyage Discovery Baffin's Bay,* app. 2, p. 62.
Astarte montagui (Dillwyn) Feyling-Hanssen, 1955: 137, pl. 21, figs. 3–12.

Pleistocene Distribution:
  Hudson Bay: 1, 3, 4, 5, 6, 7
  Labrador: 1, 4
  Quebec: 4, 7, 9
  Maine: 1, 2, 3, 4, 6, 12, 15
Present Distribution: Arctic.

Astarte elliptica (Brown), 1827

Plate 6, Figures 6, 7

Crassina sulcata Nilsson, 1822: 187. Not DeCosta, 1778.
Crassina elliptica Brown, 1827, *Illustrations of Conch. Great Britain* **1**: pl. 18, fig. 3.
Venus compressa Hanley, 1855, *Ipsa Linnaei,* London, 454.

Pleistocene Distribution:
  Labrador: 1
  Maine: 2, 15
  New York: 3
  Massachusetts: 3
Present Distribution: Arctic to Massachusetts.

For discussion of the genus *Astarte* see Dall (1903) and Ockelmann (1958).

Astarte quadrans Gould, 1841

Plate 6, Figures 8, 9

Astarte quadrans Gould, 1841, *Rept. Inv. Mass.*, 81.

Pleistocene Distribution:
  Massachusetts: 1
Present Distribution: Gulf of St. Lawrence to Long Island Sound.

Astarte borealis (Schumacker), 1817

Plate 6, Figures 10, 11

Tridonta borealis Schumacker, 1817, *Essai d'un Nouveau Système Test.*, 147, pl. 17, fig. 1.
Astarte borealis Feyling-Hanssen, 1955: 134, pl. 20, figs. 1–8, pl. 21, figs. 1–2.

Pleistocene Distribution:
  Hudson Bay: 1, 3, 4, 5, 6, 7
  James Bay: 2, 3
  Labrador: 8
  Newfoundland: 4, 16, 26
  Maine: 2
Present Distribution: Arctic seas to Massachusetts Bay; Alaska.

This is probably the *A. arctica* Möller cited by Dawson and others.

Astarte castanea (Say), 1822

Plate 6, Figures 19, 20

Venus castaneus Say, 1822, *Jour. Acad. Nat. Sci. Phila.* 4: 273.

Pleistocene Distribution:
  Massachusetts: 1, 2, 3
  New York: 5
  New Jersey: 8, 16, 17
  Maine: 5
Present Distribution: Nova Scotia to North Carolina.

Astarte laurentiana Lyell, 1845

Plate 6, Figures 17, 18

Astarte laurentiana Lyell, 1845, *Travels, N. America* 2: 150.

Pleistocene Distribution:
  Quebec: 4, 6, 9
Present Distribution: Extinct; closely related to *A. soror* Dall.

Astarte striata Leach, 1819

Plate 6, Figures 21, 22

Astarte striata Leach, 1819, *Ross's Voyage*, app. 2, p. 176.

Pleistocene Distribution:
  James Bay: 2, 3, 7, 8, 9
  Newfoundland: 2, 3, 5, 6, 10, 15, 16, 17, 38
  Labrador: 5
  Maine: 2, 4, 6
Present Distribution: Davis Strait to Massachusetts Bay.

Astarte undata Gould, 1841

Plate 6, Figure 16

Astarte undata Gould, 1841, *Rept. Invert. Massachusetts*, 80.

Pleistocene Distribution:
  Massachusetts: 1
  New Jersey: 25
Present Distribution: New Brunswick to Maine.

Astarte subaequilatera Sowerby, 1854

Plate 6, Figures 12, 13

Astarte subaequilatera Sowerby, 1854, *Thes. Conch.* 2(2): 36.
Astarte subaequilatera Dall, 1903, *Proc. U. S. Nat. Mus.* 26 (1342): 938–939.

Pleistocene Distribution:
  Maine: Deering
Present Distribution: Labrador to Florida; 22 fathoms and deeper.

This species has been referred to in the literature as *Astarte lens* Stimpson (manuscript name) and *A. crebicostata* Forbes. For discussion see Dall, 1903: 938–939.

FAMILY CYRENIDAE

Polymesoda carolinensis (Bosc), 1830

Plate 6, Figure 23

Cyclas carolinensis Bosc, 1830, *Hist. Nat. des Coquilles* 2: pl. 18, fig. 4.
Cyrena carolinensis Holmes, 1860: 31, pl. 6, fig. 7.

Pleistocene Distribution:
  North Carolina: 1
  South Carolina: 16
Present Distribution: Virginia to Texas (brackish) and northern half of Florida.

FAMILY PLEUROPHORIDAE

Arctica islandica (Linné), 1767

Plate 6, Figures 25, 26

Venus islandica Linné, 1767, *Syst. Nat.*, 12th ed., 1131.
Cyprina islandica Feyling-Hanssen, 1955: 143, pl. 22, figs. 6–9, pl. 23, figs. 1–3.

Pleistocene Distribution:
  Labrador: 7
  New York: 3
Present Distribution: Arctic to North Carolina.

FAMILY GOULDIIDAE

Crassinella lunulata (Conrad), 1834

Plate 6, Figures 27, 28

Astarte lunulata Conrad, 1834, *Jour. Acad. Nat. Sci. Phila.* 7: 133.
Astarte lunulata Holmes, 1860: 32, pl. 6, fig. 9.
Crassinella lunulata Gardner, 1943: 62, pl. 19, fig. 30.
Crassinella lunulata Olsson, 1953: 72.

Pleistocene Distribution:
  Massachusetts: 1 (after Cushman)
  Virginia: 15
  North Carolina: 1, 8, 11
  South Carolina: 12, 14, 16
Present Distribution: Massachusetts to the West Indies, both sides of Florida.

Crassinella mactracea (Linsley), 1845

Plate 6, Figure 24

Astarte mactracea Linsley, 1845, *Amer. Jour. Sci.* **48**: 275.

Pleistocene Distribution:
  Massachusetts: 1 (after Cushman)
Present Distribution: Massachusetts Bay to Long Island.

Almost identical with *C. lunulata,* but more obese, with a more oval lunule and more chalky.

FAMILY CARDITIDAE

Cardita floridana (Conrad), 1838

Plate 6, Figures 31, 32

Carditamera floridana Conrad, 1838, *Foss. Medial Tertiary,* 12.
Cardita floridana Holmes, 1860: 32, pl. 7, fig. 1.

Pleistocene Distribution:
  North Carolina: 8, 11
  South Carolina: 16
Present Distribution: South half of Florida and Gulf of Mexico.

Venericardia borealis (Conrad), 1831

Plate 6, Figures 29, 30

Cardita borealis Conrad, 1831, *Amer. Marine Conch.,* 39, pl. 8. fig. 1.

Pleistocene Distribution:
  Labrador: 1
  Newfoundland: 2, 3, 4
  Massachusetts: 1, 2
  New York: 3
  New Jersey: 8, 17, 18, 25
Present Distribution: Labrador to North Carolina (deep water off New Jersey).

Venericardia novangliae Morse, 1869

Plate 7, Figures 1, 2

Venericardia borealis novangliae Morse, 1869, *First Ann. Rept. Peabody Acad Sci.,* 76.

Pleistocene Distribution:
  Massachusetts: 1
Present Distribution: Nova Scotia to New York.

Venericardia tridentata Say, 1826

Plate 7, Figure 3

Venericardia tridentata Say, 1826, *Jour. Acad. Nat. Sci. Phila.* 5: 216.
Cardita tridentata Holmes, 1860: 31, pl. 6, fig. 8.
Glans tridentata Gardner, 1943: 70.

Pleistocene Distribution:
  New Jersey: 17
  Virginia: 7, 15
  North Carolina: 1, 8, 9, 11, 15, 17
  South Carolina: 2, 7, 9, 12, 16
Present Distribution: North Carolina to Florida and Gulf of Mexico.

FAMILY CHAMIDAE

Echinochama arcinella (Linné), 1767

Plate 7, Figure 4

Chama arcinella Linné, 1767, *Syst. Nat.,* 12th ed., 1139.
Chama arcinella Holmes, 1860: 23, pl. 5, fig. 1.
Echinochama arcinella Dall, 1903: 1415.

Pleistocene Distribution:
  South Carolina: 16
Present Distribution: North Carolina to West Indies and Brazil.

Chama macerophylla Gmelin, 1790

Plate 7, Figure 5

Chama macerophylla Gmelin, 1790, *Syst. Nat.,* 13th ed., 3305.
Chama macerophylla Dall, 1903: 1403.

Pleistocene Distribution:
  North Carolina: 8
Present Distribution: North Carolina to West Indies.

FAMILY THYASIRIDAE

Thyasira gouldi (Philippi), 1845

Plate 7, Figures 6, 7

Thyasira flexuosa Montagu, 1803, *Testacea Britannica,* 72 (in part), *fide* Feyling-Hanssen, 1955: 140.
Cryptodon gouldii Philippi, 1845, *Zeitschrift. für Malak.* 2: 74.

Pleistocene Distribution:
  Maine: 16
  Vermont: 9
  Quebec: 9
  Newfoundland: 17
Present Distribution: Labrador to off North Carolina.

FAMILY UNGULINIDAE

Diplodonta soror (Adams), 1852

Plate 7, Figures 8, 9

Lucina soror Adams, 1852, *Contrib. Conch.,* 247.
Lucina kiawahensis Holmes, 1860: 29, pl. 6, fig. 5.
Diplodonta soror Dall, 1900: 1188.
Diplodonta soror Gardner, 1943: 81, pl. 14, figs. 42, 43.

Pleistocene Distribution:
South Carolina: 14, 16
Present Distribution: North Carolina to Texas and West Indies.

Diplodonta punctata (Say), 1822

Plate 7, Figures 10, 11, 35, 36

Amphidesma punctata Say, 1822, *Jour. Acad. Nat. Sci. Phila.* 2: 308.
Diplodonta punctata Dall: 1900: 1187.

Pleistocene Distribution:
New Jersey: 17
North Carolina: 10, 11
South Carolina (after Pugh)
Present Distribution: North Carolina to both sides of Florida and the West Indies.

Diplodonta semiaspera Philippi, 1836

Plate 7, Figures 12, 13

Diplodonta semiaspera Philippi, 1836, *Arch. f. Naturg.* 2: 225, pl. 7, figs. 2a–2d.
Diplodonta semiaspera Dall, 1900: 1188.

Pleistocene Distribution:
South Carolina: 14
Present Dsitribution: North Carolina to Texas and West Indies.

Lucina filosa (Stimpson), 1851

Plate 7, Figures 34, 39

Lucina radula Gould, 1841, *Invert. of Mass.,* 69.   Not Montagu.
Lucina filosa Stimpson, 1851, *Shells of New England,* 17, no desc.
Lucinoma filosa Stimpson, Blake, 1953: 25.

Pleistocene Distribution:
Maryland: 16 (after Blake)
Present Distribution: Maine to Florida, 16 to 528 fathoms.

Phacoides crenella Dall, 1901

Plate 7, Figures 24, 25

Phacoides crenella Dall, 1901, *Proc. U. S. Nat. Mus.* 23: 810, 825, pl. 39, fig. 2.

Pleistocene Distribution:
Virginia: 16
North Carolina: 1, 6, 7, 9
South Carolina: 2, 12, 13
Present Distribution: Virginia to Cuba.

Phacoides nassula (Conrad), 1846

Plate 7, Figures 14, 15

Lucina nassula Conrad, 1846, *Amer. Jour. Sci.* 2: 394.
Phacoides nassula Dall, 1903: 1372.

Pleistocene Distribution:
North Carolina: 1
Present Distribution: North Carolina to Texas and Bahamas.

Lucina amiantus (Dall), 1901

Plate 7, Figures 18, 19

Phacoides amiantus Dall, 1901, *Synopsis Lucinacea,* 826, pl. 39, fig. 10.

Pleistocene Distribution:
North Carolina: 1
South Carolina: 14, 16
Present Distribution: North Carolina to West Indies.

Lucina multilineata Tuomey & Holmes, 1857

Plate 7, Figures 20, 21

Lucina multilineata Tuomey & Holmes, 1857, *Pleiocene Fossils of South Carolina,* 61, pl. 18, figs. 16, 17.
Lucina multilineata Holmes, 1860: 29, pl. 6, fig. 6.
Phacoides multilineatus Dall, 1903: 1384.
Phacoides multilineatus Gardner, 1943: 78, pl. 13, figs. 34–37.
Lucina multilineata Blake, 1953: 25.

Pleistocene Distribution:
Maryland: 15
North Carolina: 1
South Carolina: 14, 16
Present Distribution: North Carolina to both sides of Florida.

Phacoides radians (Conrad), 1840

Plate 7, Figures 22, 23

Lucina radians Conrad, 1840, *Trans. Amer. Assoc. Nat. & Geol.* 1: 110.
Lucina radians Holmes, 1860: 28, pl. 6, fig. 3.
Phacoides radians Dall, 1903: 1380.
Phacoides radians Olsson, 1953: 85, pl. 7, fig. 3.

Pleistocene Distribution:
North Carolina: 1
South Carolina: 7, 16
Present Distribution: North Carolina to Florida and West Indies.

Phacoides trisulcata (Conrad), 1841

Plate 7, Figures 16, 17

Lucina trisulcata Conrad, 1841, *Trans. Amer. Assoc. Nat. & Geol.* 1: 110.
Lucina trisulcata Holmes, 1860: 28, pl. 6, fig. 4.
Phacoides trisulcatus Dall, 1903: 1369.
Phacoides trisulcatus Olsson, 1953: 85, pl. 7, figs. 4, 4a, 4b.

Pleistocene Distribution:
North Carolina: 1
South Carolina: 7, 16
Present Distribution: North Carolina to Brazil.

Anodontia alba Link, 1807

Plate 7, Figures 32, 33

Anodontia alba Link, 1807, *Beschr. Natur. Samml., Rostok*, 156.
Lucina (Loripes) chrysostoma Philippi, 1847.
Anodontia alba Abbott, 1958, *Acad. Nat. Sci. Phila. Monograph*
11: 119.

Pleistocene Distribution:
North Carolina: 8, 26
Present Distribution: North Carolina to West Indies,
Florida.

Codakia costata (Orbigny), 1842

Plate 7, Figures 26, 27

Lucina costata Orbigny, 1842 (in Sagra), *Historia de la Isla de
Cuba* 5: pl. 27, figs. 40–41.
Lucina costata Holmes, 1860: 27, pl. 6, fig. 2.

Pleistocene Distribution:
North Carolina: 1
South Carolina: 12
Present Distribution: North Carolina to Brazil.

Divaricella quadrisulcata (Orbigny), 1842

Plate 7, Figures 28, 29

Lucina quadrisulcata Orbigny, 1842 (in Sagra), *Historia de la
Isla de Cuba* 5: 27, figs. 34–36.
? *Lucina divaricata* Linné, Holmes, 1860: 27, pl. 6, fig. 1.
Divaricella quadrisulcata Dall, 1903: 1389, pl. 51, fig. 1.

Pleistocene Distribution:
New Jersey: 8, 9, 12, 13, 16, 17
Maryland: 16
Virginia: 15
North Carolina: 1, 5, 6, 7, 8
South Carolina: 2, 7, 11, 12, 14, 16
Present Distribution: Massachusetts to Brazil.

FAMILY SPORTELLIDAE

Sportella constricta (Conrad), 1841

Plate 7, Figures 37, 38

Amphidesma constricta Conrad, 1841, *Amer. Jour. Sci.* 41: 347,
pl. 2, fig. 15.
Sportella constricta Dall, 1900: 1128, pl. 25, figs. 4, 4a.
Sportella constricta Gardner, 1943: 83, pl. 14, figs. 19, 20.

Pleistocene Distribution:
South Carolina: 16
Present Distribution: North Carolina to West Indies.

Sportella protexta (Conrad), 1841

Plate 7, Figures 44, 45

Amphidesma protexta Conrad, 1841, *Amer. Jour. Sci.* 41: pl. 347.
Saxicava fragilis Holmes, 1860: 57, pl. 8, fig. 18.
Sportella protexta Dall, 1900: 1129, pl. 25, fig. 3.

Pleistocene Distribution:
South Carolina: 14, 16
Present Distribution: Off North Carolina; 22 fathoms.

Anisodonta elliptica Recluz, 1850

Plate 7, Figures 42, 43

Eucharis elliptica Recluz, 1850, *Jour. de Conchyl.* 1: 168.
Mya simplex Holmes, 1860: 55, pl. 8, fig. 16.

Pleistocene Distribution:
North Carolina: 10, 11
South Carolina: 16
Present Distribution: North Carolina to West Indies.

FAMILY LEPTONIDAE

Bornia longipes (Stimpson), 1855

Plate 7, Figures 30, 31

Lepton longipes Stimpson, 1855, *Proc. Boston Soc. Nat. Hist.*
5: 111.

Pleistocene Distribution:
Virginia: 7 (after Mansfield)
Present Distribution: North and South Carolina.

Aligena elevata (Stimpson), 1851

Plate 7, Figures 40, 41

Montacuta bidentata Gould, 1841, *Invertebrata of Mass.*, 1st ed.,
59. Not of Turton, 1822.
Montacuta elevata Stimpson, 1851, *Shells of New England*, 16
(no fig.).
Aligena elevata Blake, 1953: 25.
Montacuta elevata Gould, 1870, *Invertebrata of Mass.*, 2nd ed.,
86, fig. 396.

Pleistocene Distribution:
New Jersey: 21
Maryland: 16
South Carolina: 16
Present Distribution: Massachusetts to North Caro-
lina.

Montacuta bowmani Holmes, 1860

Plate 7, Figure 46

Montacuta bowmani Holmes, 1860: 30, pl. 7, fig. 2.

Pleistocene Distribution:
South Carolina: (after Holmes)
Present Distribution: Unknown; based only on
Holmes's record. Dall (1900: 1152) refers this species
to the genus *Rochefortia*.

Mysella planulata (Stimpson), 1851

Plate 7, Figure 47

Kellia planulata Stimpson, 1851, *Shells of New England*, 17.
Rochefortia planulata Dall, 1900: 1161, pl. 45, fig. 7.
Rochefortia planulata Blake, 1953: 25.

Pleistocene Distribution:
Maryland: 15, 16 (after Blake)
North Carolina: 10, 11
Florida
Present Distribution: Nova Scotia to Texas and West
Indies.

FAMILY **CARDIIDAE**

Dinocardium robustum (Solander), 1786

Plate 8, Figures 6, 7

Cardium magnum Born, 1780, *Testacea Musei Caesarei Vindonbonensis*, 46, pl. 3, fig. 5. Not Linné, 1758.
Cardium robustum Solander, 1786, *Portland Catalogue*, 58.
Cardium magnum Born, Holmes, 1860: 23, pl. 5, figs. 2–2a. Not of Linné, 1758.
Cardium robustum Dall, 1900: 1099.
Dinocardium robustum Clench and Smith, 1944, *Johnsonia* 1 (13): 9, pl. 6.

Pleistocene Distribution:
  North Carolina: 1, 5, 9, 10, 11
  South Carolina: 7, 12, 14, 16
Present Distribution: Virginia to Northern Florida, Texas, and Mexico.

Cerastoderma pinnulatum (Conrad), 1831

Plate 8, Figures 3, 18

Cardium pinnulatum Conrad, 1831, *Jour. Acad. Nat. Sci. Phila.* 6: 260, pl. 11, fig. 8.
Cerastoderma pinnulatum Clench and Smith, 1944, *Johnsonia* 1(13): 12, pl. 8, figs. 1–7.

Pleistocene Distribution:
  New Brunswick: 1, 2, 3
  Maine: 2, 6, 15
Present Distribution: Labrador to North Carolina (deep water to the south).

Serripes groenlandicus (Bruguiere), 1789

Plate 8, Figures 12, 13

Cardium groenlandicum Bruguiere, 1789, *Encly. Méthodique*, 222.

Pleistocene Distribution:
  Hudson Bay: 10, 13
  James Bay: 2
  Labrador: 1
  Newfoundland: 1, 3, 6, 7, 8, 13, 16, 30
  Quebec: 4, 5, 7
  New Brunswick: 1
  Maine: 2, 19
Recent Distribution: Greenland to Cape Cod.

Trachycardium muricatum (Linné), 1758

Plate 8, Figures 4, 5

Cardium muricatum Linné, 1758, *Syst. Nat.*, 10th ed., 679.
Cardium muricatum Holmes, 1860: 24, pl. 5, fig. 3.
Cardium muricatum Dall, 1900: 1089.
Trachycardium muricatum Gardner, 1943: 92, pl. 15, fig. 21.
Cardium muricatum Olsson, 1953: 101, pl. 10, fig. 4.
Trachycardium (Dallocardia) muricatum Clench and Smith, 1944, *Johnsonia* 1(13): 7, pls. 1, 3.

Pleistocene Distribution:
  South Carolina: 2, 11, 14, 16
Present Distribution: North Carolina to West Indies and Gulf of Mexico.

Clinocardium ciliatum (Fabricius), 1780

Plate 8, Figures 8, 9

Cardium ciliatum Fabricius, 1780, *Fauna Groenlandica*, 410.
Cardium dawsoni Stimpson, 1862, *Proc. Acad. Nat. Sci. Phila.*, 58.
Cardium islandicum Linné, Dawson, 1872: 77.
Cardium ciliatum Dall, 1900: 1096.
Clinocardium ciliatum Feyling-Hanssen, 1955: 141, pl. 22, fig. 3.
Clinocardium ciliatum Clench and Smith, 1944, *Johnsonia* 1 (13): 15, pl. 10.

Pleistocene Distribution:
  Hudson Bay: 1, 3
  James Bay: 1, 2, 3, 5, 6
  Newfoundland: 5, 10, 15
  New Brunswick: 1, 3
  Quebec: 3, 4, 5
  Maine: 2, 10
  New Jersey: 25
Present Distribution: Arctic to Cape Cod, Alaska to Puget Sound.

Laevicardium laevigatum (Linné), 1758

Plate 8, Figures 10, 11

Cardium laevigatum Linné, 1758, *Syst. Nat.*, 10th ed., 680. Not *Cardium serratum* Linné, 1758, which is Indo-Pacific.
Laevicardium laevigatum Clench and Smith, 1944, *Johnsonia* 1(13): 22, pl. 12, figs. 1–5.
Laevicardium laevigatum Olsson, 1944: 105, pl. 11, figs 5, 5a, 5b.

Pleistocene Distribution:
  North Carolina: 1, 5
  South Carolina: 14
  Florida, Louisiana
Present Distribution: North Carolina to West Indies.

Laevicardium mortoni (Conrad), 1830

Plate 8, Figures 1, 2

Cardium mortoni Conrad, 1830, *Jour. Acad. Nat. Sci. Phila.* 6: 259, pl. 11, figs. 5, 6, 7.
Cardium mortoni Holmes, 1860: 26, pl. 5, fig. 6.
Cardium mortoni Clench and Smith, 1944, *Johnsonia* 1(13): 27, pl. 12, figs. 6, 7.

Pleistocene Distribution:
  New Jersey: 6, 8, 16, 17
  Virginia: 11
  North Carolina: 1, 5
  South Carolina: 2, 16
Present Distribution: Nova Scotia to Gulf of Mexico.

FAMILY **VENERIDAE**

Dosinia elegans Conrad, 1838

Plate 8, Figure 14

Artemis elegans Conrad, 1838, *Foss. Medial Tertiary*, 30.
Dosinia elegans Conrad, 1846, *Amer. Jour. Sci.* 2(2): 393.
Dosinia elegans Dall, 1903: 1231.

Dosinia elegans Palmer, 1927, *Paleont. Amer.* 1: 62, pl. 18,
figs. 3, 4, 8, 9; pl. 20, fig. 2.
Dosinia elegans Clench, 1942, *Johnsonia* 1(3) : 1, pl. 1.
Dosinia elegans Gardner, 1943: 122, pl. 11, fig. 1.
Dosinia elegans Olsson, 1953: 117.

Pleistocene Distribution:
    North Carolina: 1
    South Carolina: 15
Present Distribution: North Carolina to Yucatan and
West Indies.

Dosinia discus (Reeve), 1850

Plate 8, Figure 15

Artemis discus Reeve, 1850, *Conch. Icon.* 6: pl. 2, fig. 9.
Dosinia concentrica Holmes, 1860: 37, pl. 7, fig. 4.
Dosinia discus Palmer, 1927: 69, pl. 18, figs. 1, 6; 20, fig. 1.
Dosinia discus Clench, 1942, *Johnsonia* 1(3) : 4, pl. 3.

Pleistocene Distribution:
    North Carolina: 1
    Florida, Louisiana, and Texas
Present Distribution: Virginia to Vera Cruz and
Bahamas.

Pitar morrhuana (Linsley), 1845

Plate 8, Figures 16, 17

Cytherea convexa Say, 1824, *Jour. Acad. Nat. Sci. Phila.,* 1st
ser., 4 : 149, pl. 12, fig. 3. Not *C. convexa* Brongniart, 1822.
Cytherea sayana Conrad, 1833, *Amer. Jour. Sci.,* 1st ser., 23:
345 (in part).
Cytherea morrhuana Linsley, 1845, *Amer. Jour. Sci.* 48: 276.
Callocardia morrhuana Dall, 1903: 1262, pl. 54, fig. 14.
Callocardia sayana Gardner, 1943: 124, pl. 19, fig. 33.
Callocardia morrhuana Blake, 1953: 25.

Pleistocene Distribution:
    New York: 2
    New Jersey: 16, 17
    Maryland: 15, 16
    Virginia: 15
    North Carolina: 1, 26
Present Distribution: Gulf of St. Lawrence to North
Carolina.

Macrocallista nimbosa (Solander), 1786

Plate 9, Figure 1

Venus nimbosa Solander, 1786, *Portland Cat.,* 175, no. 3761.
Callista gigantea Holmes, 1860: 36, pl. 7, fig. 3.
Macrocallista gigantea Dall, 1903: 1254.
Callista (Callista) nimbosa Palmer, 1927, *Paleont. Amer.* 1: 84,
pl. 10, fig. 15; pl. 13, figs. 1–4; pl. 14, fig. 18.
Macrocallista nimbosa Olsson, 1953: 109.
Macrocallista nimbosa Clench, 1942, *Johnsonia* 1(3) : 5, pl. 4.

Pleistocene Distribution:
    North Carolina: 1, 9
    South Carolina: 16
    Florida, Texas
Present Distribution: North Carolina to Gulf of
Mexico.

Chione cancellata (Linné), 1767

Plate 9, Figure 2

Venus cancellata Linné, 1767, *Syst. Nat.,* 12th ed., 1130.
Chione cancellata Holmes, 1860: 35, pl. 6, fig. 14.
Chione cancellata Dall, 1903: 1290.
Chione (Chione) cancellata Palmer, 1927, *Paleont. Amer.* 1(5):
150, pl. 40, figs. 3, 4, 19, 29, 30; pl. 44, fig. 20.
Chione cancellata Olsson, 1953: 111.

Pleistocene Distribution:
    Virginia: 10, 11
    North Carolina: 1, 6, 7, 8, 9, 13
    South Carolina: 1, 2, 7, 12, 14, 16
Present Distribution: North Carolina to Brazil.

Chione intapurpurea (Conrad), 1849

Plate 9, Figures 3, 4

Venus intapurpurea Conrad, 1849, *Jour. Acad. Nat. Sci.,* 2nd ser.,
1: 209, pl. 39, fig. 9.
Chione (Chione) intapurpurea Palmer, 1927: 143, pl. 38, fig. 3;
pl. 39, figs. 4, 5; pl. 40, figs. 34–36.

Pleistocene Distribution:
    New Jersey: 39
    South Carolina: 13, 16
Present Distribution: North Carolina to Central
America, Gulf States and West Indies.

Close to, and possibly identical with, *Chione cribaria*
(Conrad), 1843.

Chione latilirata (Conrad), 1841

Plate 9, Figures 5, 6

Venus latilirata Conrad, 1841, *Proc. Acad. Nat. Sci. Phila.* 1: 28.
Chione latilirata Dall, 1903: 1298, pl. 42, fig. 3.
Chione (Lirophora) latilirata Palmer, 1927, *Paleont. Amer.*
1(5) : 179, pl. 41, figs. 7, 12, 13, 31–34; pl. 42, fig. 19.

Pleistocene Distribution:
    North Carolina: 1
    South Carolina: 7
Present Distribution: North Carolina to Brazil.

Chione grus (Holmes), 1858

Plate 9, Figure 7

Tapes grus Holmes, 1858, *Post-Pleiocene Foss. of South Caro-
lina,* 37, pl. 7, fig. 5.
Chione grus Dall, 1903: 1299.
Chione (Chione) grus Palmer, 1927: 156, pl. 40, figs. 13, 18, 21.
Chione grus Gardner, 1943: 128, pl. 19, figs. 12, 13, 20, 21.

Pleistocene Distribution:
    South Carolina: 14, 16
Present Distribution: North Carolina to Yucatan.

Gemma gemma (Totten), 1834

Plate 9, Figures 8, 9

Venus gemma Totten, 1834, *Amer. Jour. Sci.* **26**: 367, pl. 1, fig. 2.
Gemma gemma Dall, 1903: 1331, pl. 24, figs. 1, 3.
Gemma gemma Palmer, 1927, *Paleont. Amer.* **1**: 207, pl. 43, figs. 9, 13, 31.
Gemma gemma Blake, 1953: 25.

Pleistocene Distribution:
 Massachusetts: 1
 New York: 2, 7
 New Jersey: 7, 8, 9, 16, 17, 39
 Delaware: 3
 Maryland: 16
 Virginia: 7, 15
 North Carolina: 1, 8, 9
 Alabama
Present Distribution: Nova Scotia to Texas and the Bahamas. Puget Sound to Washington.

Mercenaria mercenaria (Linné), 1758

Plate 9, Figures 17, 18

Venus mercenaria Linné, 1758, *Syst. Nat.*, 10th ed., 686.
Mercenaria violacea Adams, Holmes, 1860: 33, pl. 6, fig. 11.
Venus mercenaria Clark, 1906: 201, pl. 58, 59.
Venus mercenaria Palmer, 1927, *Paleont. Amer.* **1**(5): 184, pl. 32, figs. 2, 3, 4, 7.
Venus mercenaria Blake, 1953: 25.

Pleistocene Distribution:
 Massachusetts: 1, 2, 3
 New York: 3, 6, 7
 New Jersey: 2, 5, 6, 7, 8, 9, 10, 11, 12, 13, 14, 16, 17, 18
 Delaware: 1, 2, 3
 Maryland: 1, 2, 15, 16
 Virginia: 5, 7, 8, 12, 13, 15
 North Carolina: 1, 3, 4, 6, 7, 8, 9, 10, 13, 15, 16
 South Carolina: 1, 2, 4, 12, 14, 16
 Georgia: 1
Present Distribution: Gulf of St. Lawrence to Gulf of Mexico.

Mercenaria mercenaria notata (Say), 1822

Plate 9, Figure 13

Venus notata Say, 1822, *Jour. Acad. Nat. Sci. Phila.* **2**: 271.
Mercenaria notata Holmes, 1860: 34, pl. 6, fig. 13.
Venus mercenaria Dall, 1903: 1312.
Venus mercenaria notata Palmer, 1917, *Paleont. Amer.* **1**(5): 186, pl. 32, fig. 5.
Venus mercenaria Gardner, 1943: 130, pl. 21, fig. 10.

Pleistocene Distribution:
 South Carolina: 16
Present Distribution: Gulf of St. Lawrence to Gulf of Mexico.

Mercenaria campechiensis (Gmelin), 1790

Plate 9, Figures 12, 14, 16

Venus campechiensis Gmelin, 1790, *Syst. Nat.*, 13th ed., **5**: 3287.
Mercenaria mortoni Holmes, 1860: 34, pl. 6, fig. 12.
Venus campechiensis Dall, 1903: 1315.
Venus campechiensis Palmer, 1927, *Paleont. Amer.* **1**(5): 187, pl. 34, figs. 1, 5, 6.

Pleistocene Distribution:
 New York: 2
 New Jersey: 5, 7, 8, 9, 11, 13, 16, 17, 18
 Maryland: 16
 Virginia: 13, 15
 North Carolina: 1, 7, 9, 21, 26
 South Carolina: 2, 12, 14, 16
Present Distribution: New Jersey to Texas and Cuba.

FAMILY **PETRICOLIDAE**

Petricola pholadiformis Lamarck, 1818

Plate 10, Figures 1–4

Petricola pholadiformis Lamarck, 1818, *Animaux sans Vert.* **5**: 505.
Petricola pholadiformis Holmes, 1860: 38, pl. 7, fig. 6.
Petricola pholadiformis Dall, 1900: 1061.
Petricola pholadiformis Clark, 1906: 201.
Petricola pholadiformis Gardner, 1943: 117.

Pleistocene Distribution:
 Massachusetts: 1
 New Jersey: 6, 7, 8, 9, 13, 16, 17, 18, 21
 Delaware: 3
 Maryland: 16
 North Carolina: 5
 South Carolina: 11, 14, 16
 Georgia: 1
Present Distribution: Gulf of St. Lawrence to Gulf of Mexico and south.

Petricola lata Dall, 1925

Plate 10, Figure 5

Petricola lata Dall, 1925, *Proc. Biol. Soc. Wash.* **38**: 90.
Petricola dactylus of authors.

Pleistocene Distribution:
 South Carolina: 16
Present Distribution: Maine to South Carolina.

Ropellaria typica (Jonas), 1844

Plate 10, Figures 6, 7

Choristodon typicum Jonas, 1844, *Zeitschr. f. Malakozool.* **1**: 185.
Petricola typica Dall, 1900: 1059.

Pleistocene Distribution:
 South Carolina: 16
Present Distribution: North Carolina to Florida and West Indies.

FAMILY **TELLINIDAE**

Tellina alternata Say, 1822

Plate 10, Figure 12

Tellina alternata Say, 1822, *Jour. Acad. Nat. Sci. Phila.* **4**: 275.
Peronaeoderma alternata Holmes, 1860: 45, pl. 8, fig. 1.
Tellina alternata Dall, 1900: 1029 (in part).
Telina alternata Olsson, 1953: 123, pl. 14, figs. 2, 3.

Pleistocene Distribution:
   North Carolina: 1
   South Carolina: 12, 13, 14, 16
   Present Distribution: North Carolina to Gulf of Mexico.

Tellina agilis Stimpson, 1859

Plate 10, Figures 10, 11

Tellina agilis Stimpson, 1858, *Amer. Jour. Sci.* **25**: 125.
Tellina tenera Say, 1822, *Jour. Acad. Nat. Sci. Phila.* **2**: 303.
   Not Schrank, 1803.
Tellina tenera Clark, 1906: 198, pl. 56, figs. 3, 6.

Pleistocene Distribution:
   Massachusetts: 1
   New Jersey: 8, 16, 17
   Maryland: 16
   Virginia: 15
   North Carolina: 1, 3
   South Carolina: 2
   Present Distribution: Gulf of St. Lawrence to Gulf of Mexico.

Tellina texana Dall, 1900

Plate 10, Figures 8, 9

Tellina polita Say, 1822, *Jour. Acad. Nat. Sci. Phila.* **2**: 276.
   Not of Spengler, 1798.
Angulus polita Holmes, 1860: 45, pl. 8, fig. 2.
Tellina sayi "Deshayes" Dall, 1900, *Trans. Wagner Free Instit. Sci.* **3** (5): 1034 (after MS name of Deshayes).
Tellina texana Dall, 1900, *Proc. U.S. Nat. Mus.* **23**: 313.

Pleistocene Distribution:
   North Carolina: 8, 10, 11
   South Carolina: 14, 16
   Present Distribution: North Carolina to Gulf of Mexico, southern half of Florida, and Cuba.

Quadrans lintea (Conrad), 1837

Plate 10, Figures 13, 14

Tellina lintea Conrad, 1937, *Jour. Acad. Nat. Sci. Phila.* **7**: 259, pl. 20, fig. 3.
Quadrans lintea Abbott, 1958, *Acad. Nat. Sci. Phila. Monograph* **11**: 135, pl. 4k, 1.

Pleistocene Distribution:
   North Carolina: 1
   South Carolina: 14, 16
   Present Distribution: North Carolina to both sides of Florida, Gulf of Mexico, and West Indies.

Macoma balthica (Linné), 1758

Plate 10, Figures 15, 16

Tellina balthica Linné, 1758, *Syst. Nat.*, 10th ed., 677.
Macoma fusca Holmes, 1860: 48, pl. 8, fig. 5.
Macoma groenlandica Beck, Dawson, 1873: 72.
Macoma balthica Dall, 1900: 1051.
Macoma balthica Clark, 1906: 199, pl. 56, figs. 7-10.

Pleistocene Distribution:
   James Bay: 2, 13, 19, 20
   Quebec: 4, 5, 7, 9, 10, 13
   New Brunswick: 1
   Ontario: 1, 2, 4
   Massachusetts: 1
   Vermont: 1, 2, 3, 4, 5, 6, 7, 8, 9, 10, 11, 12, 13
   New Jersey: 5, 8, 13
   Delaware: 3
   Maryland: 16
   Virginia: 9
   South Carolina: 11, 16
   Present Distribution: Arctic to Georgia; Bering Sea to off Monterey, California.

Macoma calcarea (Gmelin), 1790

Plate 10, Figures 17–19

Tellina calcarea Gmelin, 1790, *Linn. Syst. Nat.*, 13th ed., **1**: 3236.
Macoma calcarea Clark, 1906: 199.
Macoma calcarea Blake, 1953: 25.

Pleistocene Distribution:
   Labrador: 1
   Hudson Bay: 1, 3, 4
   James Bay: 1, 2, 4
   Newfoundland: 2, 3, 5, 6, 7, 8, 10, 15, 17, 24, 25, 26, 29, 30, 32, 35
   Quebec: 4, 5, 7, 9
   New Brunswick: 2
   Maine: 2, 19
   Vermont: 5, 9
   New Jersey: 17
   Maryland: 10, 16
   South Carolina: 11
   Greenland, Europe
   Present Distribution: Arctic to New Jersey. Bering Sea to off Monterey, California.

Macoma tenta (Say), 1834

Plate 10, Figures 22, 23

Tellina tenta Say, 1834, *Amer. Conch.*, pl. 65, fig. 3.
Peronaea tenta Holmes, 1860: 46, pl. 8, fig. 3.
Macoma tenta Dall, 1900: 1049.

Pleistocene Distribution:
   South Carolina: 16
   Present Distribution: Maine to Florida, and West Indies.

Macoma constricta (Bruguiere), 1792

Plate 10, Figures 20, 21

Solen constrictus Bruguiere, 1792, Actes Soc. d'Hist. Nat. Paris 1: 126.
Tellina cayennensis Holmes, 1860: 47, pl. 8, fig. 4.
Macoma constricta Dall, 1900: 1050.

Pleistocene Distribution:
South Carolina: 11
Present Distribution: North Carolina, Brazil, and West Indies.

Strigilla mirabilis (Philippi), 1841

Plate 10, Figures 24, 25, 36

Tellina flexuosa Say, 1822, Jour. Acad. Nat. Sci. Phila. 2: 303. Not of Montagu, 1803.
Tellina mirabilis Philippi, 1841, Arch. für Naturg. 7: 260.
Tellina flexuosa Holmes, 1860: 44, pl. 7, fig. 14.
Strigilla flexuosa Dall, 1900: 1039.
Strigilla mirabilis, Abbott, 1958, Acad. Nat. Sci. Phila. Monograph 11: 135, pl. 4, figs. i and j.

Pleistocene Distribution:
South Carolina: 14, 16
Present Distribution: North Carolina to Texas and West Indies.

Tellidora cristata (Recluz), 1842

Plate 11, Figures 3, 4, 6, 7

Lucina cristata Recluz, 1842, Rev. Zool. (Soc. Cuvier) 5: 270.
Tellidora lunulata Adams, 1858, Gen. Recent Moll. 2: 401.
Tellidora lunulata Holmes, 1860: 47, pl. 9, figs. 6, 7.
Tellidora cristata Dall, 1900: 1037.

Pleistocene Distribution:
North Carolina: 1
South Carolina: 14, 16
Present Distribution: North Carolina to Florida, Texas and West Indies.

FAMILY SEMELIDAE

Abra angulata Holmes, 1860

Plate 10, Figures 27, 28

Abra angulata Holmes, 1860: 50, pl. 8, fig. 8.
Abra angulata Dall, 1900: 998.

Pleistocene Distribution:
South Carolina: 14
Present Distribution: It is known only from the type locality, Charleston, S. C. Very close to, and possibly synonymus with, *A. aequalis* Say.

Abra aequalis (Say), 1822

Plate 10, Figure 26

Amphidesma aequalis Say, 1822, Jour. Acad. Nat. Sci. Phila. 2: 307.
Abra aequalis Holmes, 1860: 50, pl. 8, fig. 7.
Abra aequalis Dall, 1900: 998.
Abra aequalis Gardner, 1943: 104, pl. 17, figs. 12–15.
Abra aequalis Olsson, 1953: 135.

Pleistocene Distribution:
New York: 9
Virginia: 15
North Carolina: 10, 11
South Carolina: 16
Present Distribution: Connecticut to West Indies.

Semele bellastriata (Conrad), 1837

Plate 10, Figures 38, 39

Amphidesma bellastriata Conrad, 1837, Jour. Acad. Nat. Sci. Phila. 7: 239, pl. 20, fig. 4.
Semele bellastriata Dall, 1900: 993.
Semele bellastriata Gardner, 1943: 102, pl. 17, figs. 27, 28, 32, 33.
Semele bellastriata Olsson, 1953: 133.

Pleistocene Distribution:
North Carolina: 1
South Carolina: 12, 14, 16
Present Distribution: North Carolina to Gulf of Mexico and West Indies.

Semele proficua (Pulteney), 1799

Plate 11, Figures 1, 2

Tellina proficua Pulteney, 1799, Hutchin's Dorset, 29, pl. 5, fig. 4.
Semele orbiculata Holmes, 1860: 51, pl. 8, fig. 9.
Semele proficua Dall, 1900: 991.

Pleistocene Distribution:
South Carolina: 2, 11, 14, 16
Present Distribution: Virginia to Gulf of Mexico and West Indies.

Semele transversa (Say), 1830

Plate 10, Figure 37

Amphidesma transversa Say, 1830, Amer. Conch., pl. 28.
Semele transversa Holmes, 1860: 52, pl. 8, fig. 10.

Pleistocene Distribution:
South Carolina: Abapola Creek (after Holmes)
Present Distribution: ?

Cumingia tellinoides (Conrad), 1831

Plate 10, Figures 29, 30

Mactra tellinoides Conrad, 1831, Jour. Acad. Nat. Sci. Phila. 6: 258, pl. 9, figs. 2, 3.
Cumingia tellinoides Holmes, 1860: 53, pl. 8, fig. 12.
Cumingia tellinoides Dall, 1900: 1000.
Cumingia tellinoides Clark, 1906: 197, pl. 56, figs. 1, 2, 4, 5.
Cumingia tellinoides Blake, 1953: 25.

Pleistocene Distribution:
Massachusetts: 1
New Jersey: 39
Maryland: 16
Virginia: 7, 15

North Carolina: 18
South Carolina: 2, 14, 16
Present Distribution: Nova Scotia to Florida.

FAMILY DONACIDAE

Donax fossor Say, 1822

Plate 10, Figures 31, 32

Donax fossor Say, 1822, *Jour. Acad. Nat. Sci. Phila.* 2: 306.
Donax fossor Gardner, 1943: 106, pl. 23, figs. 1, 2, 10, 11.
Donax fossor Olsson, 1953: 138.

Pleistocene Distribution:
New Jersey: 8, 9, 12, 13, 16, 17
North Carolina: 1
South Carolina: 9, 11, 14, 16
Present Distribution: Long Island to Florida.

Donax variabilis Say, 1822

Plate 10, Figures 33–35

Donax variabilis Say, 1822, *Jour. Acad. Nat. Sci. Phila.* 2: 305.
Serrula variabilis Holmes, 1860: 8, fig. 6.

Pleistocene Distribution:
Virginia: 10
North Carolina: 1, 8, 9, 13, 15
South Carolina: 7, 8
Georgia: 1
Present Distribution: Virginia to Texas.

FAMILY CORBULIDAE

Corbula contracta Say, 1822

Plate 11, Figures 5, 11, 14

Corbula contracta Say, 1822, *Jour. Acad. Nat. Sci. Phila.* 2: 312.
Corbula contracta Holmes, 1860: 56, pl. 8, fig. 17.
Corbula contracta Clark, 1906: 193, pl. 53, figs. 1–4.

Pleistocene Distribution:
New Jersey: 7, 15, 16, 17
North Carolina: 1, 8, 9, 18
Present Distribution: Massachusetts to West Indies.

FAMILY SOLENIDAE

Ensis directus (Conrad), 1843

Plate 11, Figure 8

Solen ensis Conrad, 1842, *Proc. Nat. Inst. Bull.* 2: 191. Not Linné.
Solen directus Conrad, 1843, *Proc. Acad. Nat. Sci.* 1: 325.
Ensis ensis Holmes, 1860: pl. 8, fig. 13.
Ensis directus Clark, 1906: 196, pl. 55, figs. 9, 10.

Pleistocene Distribution:
Maine: 2
Massachusetts: 1, 2
New York: 2, 5, 7
New Jersey: 5, 6, 8, 9, 11, 13, 16, 17, 39
Delaware: 3
Maryland: 1, 16

Virginia: 12, 14, 15
North Carolina: 1, 5, 7, 8, 9, 10, 11, 13, 18
South Carolina: 14, 16
Florida, Louisiana
Present Distribution: Labrador to Florida.

Ensis minor Dall, 1900

Plate 11, Figures 12, 13

Ensis minor Dall, 1900, *Trans. Wagner Free Inst. Sci.* 3(5): 955.

Pleistocene Distribution:
South Carolina: 14, 16
Present Distribution: New Jersey to Florida and Texas.

Solen viridis Say, 1822

Plate 11, Figures 9, 10

Solen viridis Say, 1822, *Jour. Acad. Nat. Sci. Phila.* 2: 316.
Solen viridis Gardner, 1943: 108, pl. 23, fig. 40.

Pleistocene Distribution:
South Carolina: 16
Present Distribution: Rhode Island to northern Florida and Gulf states.

Siliqua costata (Say), 1822

Plate 12, Figure 2

Solen costatus Say, 1822, *Jour. Acad. Nat. Sci. Phila.* 2: 315.

Pleistocene Distribution:
North Carolina: 1
Present Distribution: Gulf of St. Lawrence to North Carolina.

FAMILY SANGUINOLARIIDAE

Tagelus divisus (Spengler), 1794

Plate 11, Figure 15

Solen divisus Spengler, 1794, *Skrift. Nat. Selsk.* 3: 96.
Tagelus divisus Dall, 1900: 985.
Tagelus divisus Olsson, 1953: 138.

Pleistocene Distribution:
North Carolina: 10
South Carolina: 2
Present Distribution: Massachusetts to southern Florida, Gulf states and Caribbean.

Tagelus gibbus (Spengler), 1794

Plate 11, Figure 16

Solen gibbus Spengler, 1794, *Skr. nat. Selsk.* 3: 104.
Silliquaria cariboea Holmes, 1860: 54, pl. 8, fig. 14.
Tagelus gibbus Clark, 1906: 200, pl. 57.
Tagelus gibbus Gardner, 1943: 107, pl. 22, figs. 1–4.

Pleistocene Distribution:
New York: 2
New Jersey: 5, 6, 7, 8, 11, 13, 16, 17, 18

Delaware: 1, 2, 3
Maryland: 1, 15, 16
Virginia: 9
North Carolina: 1, 3
South Carolina: 2, 11, 12, 13, 14, 16
Present Distribution: Massachusetts to Florida and
Texas.

FAMILY **MACTRIDAE**

Mactra fragilis Gmelin, 1790

Plate 11, Figures 18, 19

Mactra fragilis Gmelin, 1790, *Syst. Nat.,* 13th ed., 1(6): 3261.
    Gardner, 1943: 109, pl. 18, figs. 9–11, 13.

Pleistocene Distribution:
    South Carolina: 16
    Louisiana
Present Distribution: North Carolina to Brazil and
West Indies.

Mactra polynyma Stimpson, 1860

Plate 12, Figure 1

Mactra ovalis Gould, 1841, *Invertebrates of Mass.,* 1st ed., 53.
    Not Sowerby, 1817.
Mactra polynyma Stimpson, 1860, *Smithson Misc. Coll.* 2, art.
    6, no. 3, p. 3.

Pleistocene Distribution:
    Maine: 2, 11
Present Distribution: Hudson Bay to Cape Ann and
West Indies.

Spisula solidissima Dillwyn, 1817

Plate 11, Figures 17, 20, 21

Mactra solidissima Dillwyn, 1817, *Catalogue* 1: 140.

Pleistocene Distribution:
    Maine: 19
    Massachusetts: 1, 3
    New York: 5, 7
    New Jersey: 3, 6, 7, 8, 9, 11, 12, 15, 16
    Delaware: 3
    Maryland: 16
    Virginia: 15
    North Carolina: 1, 3, 6
    South Carolina: 7, 11, 12
Present Distribution: Labrador to Gulf of Mexico.

M. similis Say 1822 is probably an ecological variety.

Mulinia lateralis (Say), 1822

Plate 11, Figures 22, 23

Mactra lateralis Say, 1822, *Jour. Acad. Nat. Sci. Phila.* 2: 309.
Mactra lateralis Holmes, 1860: 40, pl. 7, fig. 9.

Mulinia lateralis Dall, 1898: 901.
Mulinia lateralis Olsson, 1953: 141.

Pleistocene Distribution:
    New York: 2, 3, 7
    New Jersey: 5, 8, 13, 16, 17
    Delaware: 3
    Maryland: 2, 8, 14, 15, 16
    Virginia: 7, 12, 15
    North Carolina: 1, 3, 5, 6, 7, 8, 9, 10, 11, 13, 15, 18
    South Carolina: 1, 2, 7, 8, 9, 11, 12, 13, 14, 16
    Georgia: 1
Present Distribution: New Brunswick to Texas and
West Indies.

Labiosa plicatella (Lamarck), 1818

Plate 12, Figures 7, 8

Lutraria plicatella Lamarck, 1818, *Animaux sans Vert.* 5: 470.
Lutaria canaliculata Say, 1822, *Jour. Acad. Nat. Sci. Phila.*
    2: 310.
Raeta canaliculata Holmes, 1860: pl. 7, fig. 13.
Labiosa canaliculata Dall, 1898: 907.
Raeta canaliculata Olsson, 1953: 143.

Pleistocene Distribution:
    New Jersey: 16
    North Carolina: 1
    South Carolina: 12, 16
    Florida, Louisiana, Texas
Present Distribution: New Jersey to Brazil.

Labiosa lineata (Say), 1822

Plate 12, Figure 9

Lutraria lineata Say, 1822, *Jour. Acad. Nat. Sci. Phila.* 2: 310.

Pleistocene Distribution:
    South Carolina: 16
Present Distribution: North Carolina to Brazil.

Rangia cuneata (Gray), 1831

Plate 12, Figure 16

Gnathodon cuneata Gray, 1831, *Sowerby Gen. Shells,* no. 36: 1–7.
Gnathodon cuneatus Holmes, 1860: 41, pl. 7, fig. 10.
Rangia cuneata Dall, 1898: 904.
Rangia cuneata Clark, 1906: 195, pl. 55, figs. 5–8.
Rangia cuneata Blake, 1953: 26.

Pleistocene Distribution:
    New Jersey: 17, 18 (well)
    Maryland: 7, 8, 15, 16, 17, 18
    Virginia: 1, 5
    North Carolina: 26
    South Carolina: 8, 9, 13, 18
Present Distribution: Northwest Florida to Mexico.
Rare along Atlantic coast north to Virginia.

FAMILY **MESODESMATIDAE**

Ervilia concentrica Gould, 1862

Plate 12, Figures 3, 4

Ervilia concentrica Gould, 1862, *Proc. Bost. Soc. Nat. Hist.* **8**: 280.
Ervilia concentrica Olsson, 1953: 144.
Ervilia concentrica Abbott, 1958, *Acad. Nat. Sci. Monograph* **11**: 136.

Pleistocene Distribution:
   South Carolina: 14, 16
   Florida
Present Distribution: North Carolina to both sides of Florida and West Indies.

Ervilia polita Dall, 1898

Plate 12, Figures 5, 6

Ervilia polita Dall, 1898, *Trans. Wagner Free Inst. Sci.* **3**(4): 916, pl. 33, fig. 17.
Ervilia polita Olsson, 1953: 144.

Pleistocene Distribution:
   South Carolina: 16
Present Distribution: Unknown.

Mesodesma arctatum Conrad, 1830

Plate 12, Figures 10, 13, 14

Mactra arctata Conrad, 1830, *Jour. Acad. Nat. Sci. Phila.* **6**: 257, pl. 11, fig. 1.

Pleistocene Distribution:
   Maine: 2, 19
   Massachusetts: 1
   New York: 5
   New Jersey: 17, 25
   Virginia: ?
Present Distribution: Greenland to Chesapeake Bay.

Mesodesma deaurata (Turton), 1830

Plate 12, Figures 11, 12

Mactra deauratum Turton, 1830, *Conch. Insul. Brit., Dithyra,* 71.
Mesodesma deauratum (Turton), Forbes and Hanley, 1848, *Hist. Brit. Moll.* **1**: 346.

Pleistocene Distribution:
   Quebec: 8
   Massachusetts: 1
   New Jersey: 17 ?
Present Distribution: Gulf of St. Lawrence to Georges Bank.

Mesodesma concentrica Holmes, 1860

Mesodesma concentrica Holmes, 1860: 44, pl. 6, fig. 10.

Pleistocene Distribution:
   South Carolina: 16 (after Holmes)
Present Distribution: ?

Mya arenaria Linné, 1758

Plate 12, Figure 15

Mya arenaria Linné, 1758, *Syst. Nat.,* 10th ed., 670.
Mya arenaria Holmes, 1860: 55, pl. 8, fig. 15.
Mya arenaria Clark, 1906: 194, pl. 53, figs. 5, 6; pl. 54, figs. 1–4.
Mya arenaria Gardner, 1943: 138, pl. 19, figs. 31, 32.
Mya arenaria Foster, 1946, *Johnsonia* **2**(20): 32, pls. 20, 21.
Mya arenaria Blake, 1953: 26.

Pleistocene Distribution:
   Hudson Bay: 3
   Newfoundland: 1, 5, 7, 11, 26, 38
   Labrador: 1
   Quebec: 5, 7, 9, 12
   New Brunswick: 1, 2, 3, 8
   Maine: 1, 2, 6, 7, 8, 13, 16, 18, 19, 20
   Massachusetts: 1, 3
   Vermont: 1, 2, 7
   New York: 2, 3
   New Jersey: 2, 5, 6, 16, 17
   Delaware: 1
   Maryland: 1, 15, 16
   North Carolina: 5
   South Carolina: 13, 16
Present Distribution: Arctic to Florida. Introduced to western United States.

Mya truncata Linné, 1758

Plate 13, Figures 1, 2

Mya truncata Linné, 1758, *Syst. Nat.,* 10th ed., 670.
Mya truncata Feyling-Hanssen, 1955: 148, pl. 25.
Mya truncata Foster, 1946, *Johnsonia* **2**(20): 30, pls. 17–19.

Pleistocene Distribution:
   Hudson Bay: 1, 3, 7
   James Bay: 2
   Newfoundland: 2, 3, 4, 6, 7, 8, 10, 15, 16, 17, 19, 24, 25, 26, 29, 30, 32, 35, 36
   Labrador: 1
   Quebec: 4, 7, 9
   New Brunswick: 3
   Maine: 2, 4, 6, 7, 13, 19
   Massachusetts: 1
Present Distribution: Arctic to Massachusetts. Arctic to Washington.

FAMILY **SAXICAVIDAE**

Cyrtodaria siliqua (Spengler), 1793

Plate 13, Figures 3, 4

Mya siliqua Spengler, 1793, *Skr. nat. Selsk.* (Copenhagen) **3**(1): 48.

Pleistocene Distribution:
   Labrador: 2
   Quebec: 4
   Newfoundland: 16
Present Distribution: Arctic to Rhode Island.

Panope arctica (Lamarck), 1818

Plate 13, Figures 5, 6

Glycymeris arctica Lamarck, 1818, *Animaux sans Vert.* **5** : 458.
Mya norvegica Spengler, 1793. Not Gmelin.

Pleistocene Distribution:
   Labrador: 2
   Quebec: 4
   Massachusetts: 1
Present Distribution: Arctic Ocean to Georges Bank;
25 to 115 fathoms; circumpolar.

Hiatella arctica (Linné), 1767

Plate 12, Figures 17–20

Mya arctica Linné, 1767, *Syst. Nat.*, 12th ed., 1113.
Saxicava arctica of authors.

Pleistocene Distribution:
   Hudson Bay: 1, 3, 4, 5
   James Bay: 1, 2, 3
   Labrador: 1
   Newfoundland: 3, 5, 6, 7, 8, 11, 15, 16, 17, 18, 20,
      22, 25, 26, 29, 30, 37, 38
   Quebec: 2, 4, 5, 7, 9, 11, 18
   Ontario: 4
   New Brunswick: 1, 7
   Maine: 2, 4, 6, 10, 13, 16, 19
   Vermont: 1, 2, 3, 4, 5, 6, 9, 12
   Massachusetts: 1
   New Jersey: 16, 17, 25
   Virginia: 15
   North Carolina: 1, 6
Present Distribution: Arctic to West Indies in deep
water. Arctic to off Panama in deep water.

Hiatella Daudin 1801 replaces the better known
Saxicava Bellevue 1802. The species is very variable
and has received several names (*rugosa, pholadis,* etc.)
but they are probably all a single species *arctica*. The
specimens obtained in deep water off the South Atlantic
coast and from Pleistocene deposits in New Jersey,
Virginia, and North Carolina are small with a thin shell.

Family PHOLADIDAE

Zirphaea crispata (Linné), 1758

Plate 13, Figure 11

Pholas crispata Linné, 1758, *Syst. Nat.*, 10th ed., 670.
Zirphaea crispata Feyling-Hanssen, 1955: 150, pl. 24, figs. 6, 7.
Zirphaea crispata Turner, 1954, *Johnsonia* 3(33) : 55, pls. 1, 3,
   28, 29, 30.

Pleistocene Distribution:
   Maine: 2, 10, 15, 19
Present Distribution: Newfoundland to New Jersey.

Gastrochaena cuneiformis Spengler, 1783

Plate 13, Figures 7, 8

Gastrochaena cuneiformis Spengler, 1783, *Nye Saml. K. Danske
   Skrifter* 2 : 180.

Pleistocene Distribution:
   South Carolina: 16
Present Distribution: North Carolina to West Indies.

Barnea costata (Linné), 1758

Plate 13, Figure 9

Pholas costatus Linné, 1758, *Syst. Nat.*, 10th ed., 669.
Pholas costata Holmes, 1860: 58, pl. 9, figs. 1, 1a.
Barnea costata Clark, 1906: 192, pl. 52.
Barnea costata Olsson, 1953: 152.
Barnea costata Blake, 1953: 26.
Cyrtopleura (Scobinopholas) costata Turner, 1954, *Johnsonia*
   3(33) : 35, pls. 17, 18.

Pleistocene Distribution:
   Delaware: 3
   Maryland: 1, 14, 15, 16
   Virginia 1, 11
   North Carolina: 1, 5, 9
   South Carolina: 2, 11, 12, 16
Present Distribution: Massachusetts to Texas and
West Indies.

Pholas campechiensis Gmelin, 1790

Plate 13, Figure 10

Pholas campechiensis Gmelin, 1790, *Syst. Nat.*, 13th ed., **6** : 3216.
Pholas campechiensis Turner, 1954, *Johnsonia* 3(33) : 48, pls.
   24, 25.

Pleistocene Distribution:
   North Carolina: 26
   South Carolina: 16
Present Distribution: North Carolina to Central
America and West Indies.

Pholas truncata Say, 1822

Plate 13, Figure 14

Pholas truncata Say, 1822, *Jour. Acad. Nat. Sci. Phila* 2: 321.
Pholas truncata Holmes, 1860: 57, pl. 9, fig. 4.
Barnea truncata Blake, 1953: 26.
Barnea (Anchomasa) truncata Turner, 1954, *Johnsonia* 3: 27,
   pls. 8, 11, 13.

Pleistocene Distribution:
   New Jersey: 9, 12, 13, 16, 17
   Maryland: 16
   South Carolina: 16
Present Distribution: Maine to Gulf of Mexico and
South Florida.

Martesia cunieformis (Say), 1822

Plate 13, Figures 12, 13

Pholas cunieformis Say, 1822, *Jour. Acad. Nat. Sci. Phila.* 2: 322.
Pholas cunieformis Holmes, 1860: 59, pl. 9, fig. 3.
Pholas cunieformis Blake, 1953: 26.

Pleistocene Distribution:
    Maryland: 16
    South Carolina: 14, 16
Present Distribution: Connecticut to West Indies.

Bankia gouldi (Bartsch), 1908

Plate 13, Figures 16, 17

Xylotrya gouldi Bartsch, 1908, *Proc. Biol. Soc. Wash.* 21: 211–212.
Bankia gouldi Blake, 1953: 26.
Bankia gouldi Clench and Turner, 1946, *Johnsonia* 2(19): 13, pl. 9, figs. 1–4.

Pleistocene Distribution:
    Maryland: 16
Present Distribution: Virginia to Texas.

Xylotrya palmulata (Lamarck), 1819

Plate 13, Figure 15

Teredo palmulatus Lamarck, 1819, *Animaux sans Vert.*, 2nd ed., 38.
Xylotrya palmulata Holmes, 1860: 60, pl. 9, fig. 5.

Pleistocene Distribution:
    South Carolina: 13
Present Distribution: ?

FAMILY TEREDIDAE

Teredo sp.

Pleistocene Distribution:
    North Carolina: 25
Present Distribution: World-wide.

X. GASTROPODA

FAMILY ACMAEIDAE

Acmaea testudinalis (Müller), 1776

Plate 14, Figures 1, 2

Patella testudinalis Müller, 1776, *Z. Dan. Prod.*, 237.

Pleistocene Distribution:
    Hudson Bay: 4, 6
    James Bay: 2, 3
    Newfoundland: 3
    Labrador: 1
Present Distribution: Arctic Seas to Connecticut.

FAMILY LEPETIDAE

Lepeta caeca (Müller), 1776

Plate 14, Figures 3, 4

Patella caeca Müller, 1776, *Z. Dan. Prod.*, 237.
Lepeta caeca Feyling-Hanssen, 1955: 154, pl. 24, fig. 13.

Pleistocene Distribution:
    Quebec: 4, 7, 19
    Labrador: 1
Present Distribution: Arctic to Massachusetts.

FAMILY FISSURELLIDAE

Diodora cayenensis (Lamarck), 1822

Plate 14, Figure 5

Fissurella cayenensis Lamarck, 1822, *Animaux sans Vert.* 6(2): 12.
Fissurella alternata Say, 1822, *Jour. Acad. Nat. Sci. Phila.* 2: 244.
Fissurediea alternata Dall, 1892: 428.
Diodora alternata alternata Johnson, 1934: 67.
Diodora cayenensis Farfante, 1943, *Johnsonia* 1(11): 5, pl. 2, figs. 1–6.

Pleistocene Distribution:
    New Jersey: Beach wash
    South Carolina: 11, 12, 14, 16
Present Distribution: North Carolina to West Indies.

Puncturella noachina (Linné), 1771

Plate 14, Figure 7

Patella noachina Linné, 1771, *Mant.*, 551.
Cemoria princeps Mighels and Adams, *Proc. Boston Soc. Nat. Hist.* 4: 42.

Pleistocene Distribution:
    Hudson Bay: 4
    Quebec: 4, 7
    Massachusetts: 1
    Maine: 3, 9
    New Jersey: 25
Present Distribution: Circumpolar to Connecticut.

FAMILY TROCHIDAE

Solariella obscura (Couthouy), 1838

Plate 14, Figures 8, 9

Turbo obscurus Couthouy, 1838, *Boston Jour. Nat. Hist.* 2: 100.
Solariella obscura Abbott, 1954: 110, fig. 31f, g.

Pleistocene Distribution:
    Massachusetts: 1
Present Distribution: Labrador to off Chesapeake Bay.

Margarites olivacea Brown, 1827

Plate 14, Figure 12

Margarites olivaceus Brown, 1827, *Illus. Conch. Great Britain and Ireland*, pl. 46, figs. 30, 31.
Margaritta argentata Gould, 1841, *Rep. Invert. Mass.*, 256.

Pleistocene Distribution:
Quebec: 4, 9 (after Dawson, rare)
Present Distribution: Labrador.

Margarites cinerea (Couthouy), 1838

Plate 14, Figures 10, 11

Turbo cinerea Couthouy, 1838, *Boston Jour. Nat. Hist.* **3**: 99, pl. 3, fig. 9.
Margarites cinereus Feyling-Hanssen, 1955: 156.
Margarites costalis Gould, Abbott, 1954, *American Seashells*, 107, pl. 17.

Pleistocene Distribution:
Labrador: 5
Newfoundland: 2
Quebec: 4
Maine: 2, 19
Present Distribution: Greenland to Massachusetts Bay.

Margarites helicina (Phipps), 1774

Plate 14, Figures 13, 14

Turbo helicina Phipps, 1774, *Voyage to the North Pole*, app., 198.
Margarites helicinus Feyling-Hanssen, 1955: 155.

Pleistocene Distribution:
Quebec: 5, 9 (after Dawson)
Present Distribution: Greenland to Massachusetts.

Margarites umbilicalis Broderip and Sowerby, 1829

Margarites umbilicalis Broderip and Sowerby, 1829, *Malac. and Conchol. Mag.* **1**: 26.

Pleistocene Distribution:
Hudson Bay: 6
Present Distribution: Labrador; Greenland; circumpolar.

FAMILY CYCLOSTREMATIDAE

Teinostoma cryptospira (Verrill), 1884

Plate 14, Figures 15, 16

Rotella cryptospira Verrill, 1884, *Trans. Conn. Acad. Sci. and Arts* **6**: 241.
Teinostoma cryptospira Dall, 1892: 414.
Teinostoma cryptospira Bush, 1900, *Trans. Conn. Acad. Sci.* **10**: 118.

Pleistocene Distribution:
Maryland: 16
Present Distribution: Off North Carolina and Florida; 30–150 fathoms.

Vitrinella multicarinata Dall, 1889

? *Architectonica gemma* Holmes, 1860: 92, pl. 14, figs. 6, 6a, 6b.
Vitrinella multicarinata Dall, 1889, *Rept. Blake Gastropoda*, 273, 361, 392.
Vitrinella multicarinata Dall, 1892: 419.

Pleistocene Distribution:
South Carolina: 13
Present Distribution: Off North Carolina; 15 fathoms.

Cochliolepis nautiliformis (Holmes), 1860

Plate 14, Figure 36

Adeorbis nautiliformis Holmes, 1860: 93, pl. 14, figs. 8, 8a, 8b.
Adeorbis nautiliformis Dall, 1892: 419.
A. nautiliformis Olsson *et al.*, 1953: 432, pl. 52, figs. 3–3e.

Pleistocene Distribution:
South Carolina: 14
Present Distribution: Florida.

Adeorbis holmesii (Dall), 1889

Plate 14, Figures 18, 19

Cochliolepis parasiticus Holmes, 1860: 93, pl. 14, fig. 9. Not of Stimpson, 1858.
Vitrinella holmesii Dall, 1889, *Rept. Blake Gastropoda*, 360, 392.
Adeorbis holmesii Dall, 1892: 346.

Pleistocene Distribution:
North Carolina: 10
South Carolina: 16
Present Distribution: Unknown.

FAMILY EPITONIIDAE

Epitonium rupicolum (Kurtz), 1860

Plate 14, Figures 20, 21

Scalaria lineata Say, 1822, *Jour. Acad. Nat. Sci. Phila.* **2**: 242. Not *Epitonium lineatum* Röding, 1798.
Scalaria rupicola Kurtz, 1860, *Catalogue of the Recent Shells Found on the Coast of North and South America*, Portland, Maine, 7.
Scalaria lineata Holmes, 1860: 90, pl. 14, fig. 3.
Scala lineata Dall, 1890: 158.
Scala lineata Clark, 1906: 185, pl. 50, figs. 1, 2.
Epitonium rupicolum Clench and Turner, 1951, *Johnsonia* **2**(30).

Pleistocene Distribution:
New Jersey: 3, 12, 13, 16
Maryland: 15, 16
Virginia: 15
North Carolina: 1, 9, 10
South Carolina: 14, 16
Present Distribution: Massachusetts to Texas.

Epitonium lamellosum (Lamarck), 1822

Plate 14, Figure 22; Plate 16, Figure 2

Scalaria lamellosa Lamarck, 1822, *Animaux sans Vert.* **6**(2): 227.
Scalaria clathrus of authors. Not of Linné.
Epitonium (Gyroscala) lamellosum Clench and Turner, 1951, *Johnsonia* **2**(30): 281, pls. 128, 129.

Pleistocene Distribution:
South Carolina: 16 (after Holmes)
Present Distribution: Southern Florida to Caribbean.

Epitonium humphreysii (Kiener), 1845

Plate 14, Figures 23, 24

Scalaria humphreysii Kiener, 1845, Iconographie des Coquilles Vivantes 10: 15, pl. 5, fig. 16.
Epitonium humphreysii Clench and Turner, 1951, Johnsonia 2(20): 268, pl. 117, fig. 2; pls. 119, 120.

Pleistocene Distribution:
New Jersey: 9, 12, 13, 16
Maryland: 16
North Carolina: 1
South Carolina: 14, 16
Present Distribution: Massachusetts to Texas.

Epitonium multistriatum (Say), 1826

Plate 14, Figure 25

Scalaria multistriata Say, 1826, Jour. Acad. Nat. Sci. Phila. 5: 208.
Scalaria multistriatum Holmes, 1860: 90, pl. 14, fig. 4.
Epitonium (Asperiscala) multistriatum Clench and Turner, 1952, Johnsonia 2(31): 292, pls. 133, 134.

Pleistocene Distribution:
Maryland: 16
North Carolina: 18
South Carolina: 16
Present Distribution: Massachusetts to Florida.

Epitonium angulatum (Say), 1830

Plate 14, Figure 26

Scala clathrus angulata Say, 1830, Amer. Conchol. 3: pl. 27.
Scalaria angulata Holmes, 1860: 89, pl. 14, fig. 2.
Epitonium angulatum Clench and Turner, 1951, Johnsonia 2 (30): 271, pl. 121, figs. 1-3; pl. 122, figs. 1-4.

Pleistocene Distribution:
Maryland: 15
South Carolina: 14, 16
Present Distribution: Connecticut to Texas.

Epitonium denticulatum (Sowerby), 1844

Plate 14, Figure 27

Scalaria denticulata Sowerby, 1844, Thes. Conch. 1: 87, pl. 32, figs. 25-26.
Epitonium angulatum Blake, 1953: 27.
Epitonium (Asperiscala) denticulatum Clench and Turner, 1952, Johnsonia 2(31): 310, pls. 147, 148.

Pleistocene Distribution:
Maryland: 16
Present Distribution: Fort Pierce, Florida, to Bahamas and Virgin Islands.

Epitonium greenlandicum (Perry), 1811

Plate 14, Figure 28

Scalaria greenlandica Perry, 1811, Conchology, pl. 28, fig. 8.
Epitonium greenlandicum Clench and Turner, 1952, Johnsonia 2(31): 320, pl. 154.

Pleistocene Distribution:
Quebec: 4
Maine
Massachusetts: 1
Present Distribution: Greenland to Rhode Island (deep water).

Acirsa costulata (Mighels and Adams), 1842

Plate 14, Figures 29-31

Turritella costulata Mighels and Adams, 1842, Jour. Boston Soc. Nat. Hist. 4: 50, pl. 4, fig. 20.

Pleistocene Distribution:
Quebec: 4, 6
Present Distribution: Off Maine; 10-40 fathoms.

Family MELANELLIDAE

Melanella intermedia (Cantraine), 1835

Plate 14, Figure 34

Eulima intermedia Cantraine, 1835, Bull. Acad. Roy. Bruxelles 2: 390.
Eulima intermedia Jeffreys, 1869, Brit. Conch: 214, pl. 77, fig. 4.
Eulima intermedia Dall, 1890: 159.

Pleistocene Distribution:
North Jersey: 13
North Carolina: 1
Present Distribution: Vineyard Sound to Georgia.

Melanella conoidea (Kurtz and Stimpson), 1851

Plate 14, Figures 32-33

Eulima conoidea Kurtz and Stimpson, 1851, Proc. Boston Soc. Nat. Hist. 4: 115.
Eulima conoidea Dall, 1890: 159, pl. 5, fig. 11.

Pleistocene Distribution:
South Carolina: 14, 16
Present Distribution: Massachusetts to Gulf of Mexico.

Family PYRAMIDELLIDAE

Pyramidella crenulata (Holmes), 1860

Plate 14, Figure 35

Obeliscus crenulatus Holmes, 1860, Post-Pleiocene Fossils of South Carolina, 88, pl. 13, figs. 14, 14a.
Pyramidella crenulata Dall, 1892: 247.

Pleistocene Distribution:
North Carolina: 5, 10, 11
South Carolina: 14, 16
Present Distribution: South Carolina to Florida and West Indies.

Turbonilla Risso, 1826

The species of this genus were not studied in the present investigation. The following records are taken from the literature:

Turbonilla interrupta Totten: Nantucket; New Jersey;
Maryland; North Carolina; South Carolina.
Turbonilla conradi Bush: New Jersey; North Carolina.
Turbonilla nivea Stimpson: South Carolina.
Turbonilla reticulata Adams: Virginia; North Caro-
lina; South Carolina.
Turbonilla speira Ravenel: South Carolina.
Turbonilla protracta Dall: South Carolina.
Turbonilla pupoides Orbigny: South Carolina.
Turbonilla puncta Adams: Virginia.

Odostomia Fleming, 1817

The species of this genus were not studied in the
present investigation. The following records are taken
from the literature:

Odostomia impressa Say: Nantucket; Maryland; Vir-
ginia; South Carolina.
Odostomia impressa granitina Dall: Hudson River
Tunnel.
Odostomia seminuda Adams: Nantucket; Maryland;
North Carolina; South Carolina.
Odostomia trifida Totten: Nantucket.
Odostomia bisuturalis Say: Nantucket.
Odostomia fusca Adams: Nantucket; Maryland.
Odostomia acutidens Dall: Maryland.
Odostomia melanoides (Conrad): Maryland.
Odostomia disparilis Verrill: Maryland.
Odostomia cf. *hendersoni* Bartsch: Maryland.

FAMILY NATICIDAE

Natica clausa Broderip and Sowerby, 1829

Plate 15, Figures 6, 7

? *Natica affinis* Gmelin, 1792, *Syst. Nat.*, 13th ed., 1 (6): 3675.
Natica clausa Broderip and Sowerby, 1829, *Zool. Jour.* 4: 372.
Natica clausa Feyling-Hanssen, 1955: 162, pl. 26, fig. 13.

Pleistocene Distribution:
Quebec: 4, 7, 9
Labrador: 1
New Brunswick: 1, 3
Maine: 2, 3, 6, 15, 19
Present Distribution: Arctic to North Carolina.

Natica pusila Say, 1822

Plate 15, Figures 8, 9

Natica pusila Say, 1822, *Jour. Acad. Nat. Sci. Phila.* 2: 257.
Natica pusila Holmes, 1860: 80, pl. 12, figs. 15, 15a.
Natica pusila Dall, 1892: 367.

Pleistocene Distribution:
New Jersey: 13, 16, 25
South Carolina: 14, 16
Present Distribution: Cape Cod to Gulf states and
West Indies.

Polinices duplicatus (Say), 1822

Plate 15, Figure 1

Natica duplicata Say, 1822, *Jour. Acad. Nat. Sci. Phila.* 2: 247.
Natica duplicata Holmes, 1860: 80, pl. 12, fig. 14.
Natica duplicata Dall, 1892: 368.
Polinices (Neverita) duplicatus Clark, 1906: 191, pl. 51, figs. 9, 10.
Polinices (Neverita) duplicata Olsson *et al.*, 1953: 268, pl. 57, fig. 3.

Pleistocene Distribution:
Massachusetts: 1
New York: 2, 3, 7
New Jersey: 3, 7, 11, 12, 13, 16, 17
Delaware: 3
Maryland: 1, 15, 16
Virginia: 12, 15
North Carolina: 1, 6, 7, 8, 9, 10, 11, 18
South Carolina: 2, 7, 9, 11, 12, 14, 16
Georgia: 2
Present Distribution: Massachusetts to the Gulf of
Mexico.

Polinices heros (Say), 1822

Plate 15, Figures 2–4, 10

Natica heros Say, 1822, *Jour. Acad. Nat. Sci. Phila.* 2: 248.
Polynices (Lunatia) heros, Dall, 1892: 373.

Pleistocene Distribution:
Quebec: 6
Massachusetts: 1
New York: 3, 7
New Jersey: 8, 9, 16, 17
Virginia: 16
North Carolina: 1
South Carolina: 13
Present Distribution: Gulf of St. Lawrence to North
Carolina.

Polinices triseriatus (Say), 1826

Plate 15, Figure 29

Natica triseriata Say, 1826, *Jour. Acad. Nat. Sci. Phila.*, 211.

Pleistocene Distribution:
Massachusetts: 1
New Jersey: 9, 16
Present Distribution: Gulf of St. Lawrence to North
Carolina.

Polinices groenlandicus (Beck) (Möller), 1842

Plate 15, Figure 5

Natica groenlandica Beck, Möller, 1842, *Index Moll. Groenl.*, 7.

Pleistocene Distribution:
Labrador: 1
James Bay: 1, 2
Newfoundland: 3, 5, 7, 8, 15, 16, 17, 26, 30

Quebec: 4, 7, 9
Maine: 2, 6, 7, 19, 20
Present Distribution: Greenland to North Carolina;
70–80 fathoms.

Sinum perspectivum (Say), 1831

Plate 15, Figure 12

Sigaretus perspectivus Say, 1831, *Amer. Conchol.*, 3, pl. 25.
Catinus perspectivus Holmes, 1860: 81, pl. 12, fig. 16.
Sinum perspectivum, Olsson *et al.*, 1953: 272, pl. 47, fig. 5.

Pleistocene Distribution:
  New Jersey: 16, 17
  North Carolina: 1, 8, 10
  South Carolina: 14, 16
Present Distribution: Virginia to the Gulf states and
West Indies.

Family LAMELLARIIDAE

Velutina undata Brown, 1839

Plate 15, Figure 11

Velutina undata Brown, 1839, *Mem. Wernerian Nat. Hist. Soc.*
  8: 102, pl. 1, fig. 15.
Velutina zonata Gould, 1841, *Rept. Invert. Mass.*, 22.

Pleistocene Distribution:
  Quebec: 6
Present Distribution: Gulf of St. Lawrence to Cape
Cod.

Capulus ungaricus Linné, 1767

Plate 16, Figure 6

Capulus ungaricus Linné, 1767, *Syst. Nat.*, 12th ed., 1259.

Pleistocene Distribution:
  Quebec: 9 (Redpath Museum)
Present Distribution: Greenland to Florida; 1 to 458
fathoms.

Family CALYPTRAEIDAE

Crucibulum striatum (Say), 1826

Plate 15, Figures 13–16, 24, 25

Calyptraea striatum Say, 1826, *Jour. Acad. Nat. Sci. Phila.* **5**:
  216.
Calyptraea striatum, Dall, 1892: 351.

Pleistocene Distribution:
  Massachusetts: 1
  New Jersey: 17, 25
Present Distribution: Nova Scotia to Florida Keys;
3–189 fathoms.

Calyptraea centralis (Conrad), 1841

Plate 15, Figures 26–27

Infundibulum centralis Conrad, 1841, *Amer. Jour. Sci.* 41: 348.
Calyptraea centralis Olsson *et al.*, 1953: 277.

Pleistocene Distribution:
  South Carolina: 16
Present Distribution: North Carolina to Texas and
the West Indies.

Family CREPIDULIDAE

Crepidula fornicata (Linné), 1758

Plate 15, Figures 21, 22

Patella fornicata Linné, 1758, *Syst. Nat.*, 10th ed., 781.
Crypta fornicata Holmes, 1860: 95, pl. 14, fig. 11.
Crepidula fornicata Dall, 1892: 356.
Crepidula fornicata Clark, 1906: 189, pl. 51, figs. 1–4.
Crepidula fornicata Olsson *et al.*, 1953: 277.

Pleistocene Distribution:
  Massachusetts: 1
  New York: 2, 3, 6
  New Jersey: 3, 8, 9, 11, 12, 13, 16, 17
  Delaware: 1, 2, 3
  Maryland: 1, 2, 15, 16
  Virginia: 7, 13
  North Carolina: 1, 7, 8, 10, 18, 19, 24, 25
  South Carolina: 2, 14, 16
Present Distribution: Prince Edward Island to Texas
and West Indies.

Crepidula plana Say, 1822

Plate 15, Figure 17

Crepidula plana Say, 1822, *Jour. Acad. Nat. Sci. Phila.* 2: 226.
Crepidula plana Dall, 1892: 358.
Crepidula plana Clark, 1906: 190, pl. 51, figs. 5–8.
Crepidula plana Olsson *et al.*, 1953: 279.

Pleistocene Distribution:
  Massachusetts: 1
  New York: 2, 3
  New Jersey: 3, 8, 9, 11, 12, 13, 16, 17
  Delaware: 1
  Maryland: 1, 15, 16
  Virginia: 15
  North Carolina: 1, 8, 10, 13, 18
  South Carolina: 4, 16
Present Distribution: Prince Edward Island to Texas.

Crepidula convexa Say, 1822

Plate 15, Figure 23

Crepidula convexa Say, 1822, *Jour. Acad. Nat. Sci. Phila.* 2:
  227.
Crepidula convexa Dall, 1892: 357.

Pleistocene Distribution:
  Massachusetts: 1
  New York: 2
  New Jersey: 3, 5, 7, 8, 11, 12, 13, 16, 17
  Maryland: 20
  Virginia: 15

North Carolina: 1, 3, 6, 9, 13
South Carolina: 4, 16
Present Distribution: Nova Scotia to Texas.

Crepidula aculeata (Gmelin), 1791

Plate 15, Figure 18

Patella aculeata Gmelin, 1791, *Syst. Nat.,* 13th ed., 3693.
Crypta aculeata Holmes, 1890: 95, pl. 14, fig. 12.
Crepidula (Bostrycapulus) aculeata Olsson et al., 1953: 280.

Pleistocene Distribution:
South Carolina: 14 ?
Present Distribution: North Carolina to Texas and West Indies.

FAMILY ARCHITECTONICIDAE

Architectonica nobilis Röding, 1791

Plate 15, Figures 19, 20

Architectonica nobilis Röding, 1791, *Museum Bolton,* 78, no. 1025.
Solarium granulatum Lamarck, 1792, *Animaux sans Vert.* 7: 3.
Solarium granulatum Dall, 1890: 329.
Architectonica nobilis Abbott, 1954, *American Seashells,* 142, pl. 4m.

Pleistocene Distribution:
South Carolina: 16
Present Distribution: North Carolina to Gulf of Mexico and the West Indies.

"Angaria" crassa Holmes, 1860

Plate 15, Figure 28

Angaria crassa Holmes, 1860: 92, pl. 14, figs, 7, 7a, 7b.

Pleistocene Distribution:
South Carolina: 14, 16 (after Holmes)
Present Distribution: ?

FAMILY LITTORINIDAE

Littorina irrorata (Say), 1822

Plate 15, Figure 32

Turbo irroratus Say, 1822, *Jour. Acad. Nat. Sci. Phila.* 2: 239.
Littorina irrorata Holmes, 1860: 91, pl. 14, fig. 5.
Littorina irrorata Dall, 1892: 320.
Littorina irrorata Bequaert, 1943, *Johnsonia* 1 (7): 6, pl. 2, figs. 1-7.
Littorina irrorata Olsson et al., 1953: 328.

Pleistocene Distribution:
Connecticut: 1
New Jersey: 12, 13, 17
Maryland: 1
Virginia: 8, 11
North Carolina: 1, 5
South Carolina: 11, 12, 14, 16
Georgia: 2
Present Distribution: Massachusetts (?) to northern Florida to Texas, Gulf of Mexico.

Littorina saxatilis (Olivi), 1792

Plate 15, Figure 30

Turbo saxatilis Olivi, 1792, *Zool. Adriatica,* 172, pl. 5, figs. 3a–d.
Turbo rudis Maton, 1797, *Observations Nat. Hist. Antiq. Western Counties,* 277 (Devon, England).
Turbo rudis Donovan, 1800, *Brit. Shells.* 1: pl. 33, fig. 3.
Littorina saxatilis Bequaert, 1943, *Johnsonia* 1 (7): 8, pl. 3, figs. 1-10.
Littorina saxitilis Feyling-Hanssen, 1955: 159.

Pleistocene Distribution:
Hudson Bay: 4
James Bay: 2, 3 ?
Newfoundland: 17, 38
Quebec: 4
Present Distribution: Arctic seas to New Jersey.

Turritella (Turritellopsis) acicula Stimpson, 1851

Plate 15, Figure 34

Turritella acicula Stimpson, 1851, *Proc. Boston Soc. Nat. Hist.* 4: 15, *Shells of New England,* 35, pl. 1, fig. 5.

Pleistocene Distribution:
Quebec: 4 (after Dawson) ?
Labrador: 2
Present Distribution: Labrador to Massachusetts Bay.

Distinguished from the young *T. erosa* by the more convex whorls and prominent ribs.

Turritella (Tachyrhynchus) erosa Couthouy, 1838

Plate 15, Figure 33

Turritella erosa Couthouy, 1838, *Boston Jour. Nat. Hist.* 2: 103.

Pleistocene Distribution:
Hudson Bay: 1, 6
Labrador: 1
Quebec: 4, 9
Present Distribution: Arctic to Massachusetts Bay.

FAMILY TURRITELLIDAE

Turritella (Tachyrhynchus) reticulata Mighels and Adams, 1842

Plate 15, Figure 31; Plate 16, Figure 1

Turritella reticulata Mighels and Adams, 1842, *Boston Jour. Nat. Hist.* 4: 50.

Pleistocene Distribution:
Labrador: (after Packard and Dawson) 2, 5
Present Distribution: Greenland to Gulf of St. Lawrence.

Vermicularia spirata (Philippi), 1836

Plate 16, Figures 8, 9

Vermiculus spiratus Philippi, 1836, *Arch. f. Naturg.* 2: 244.
Serpulorbis spirata Dall, 1892: 304.

Pleistocene Distribution:
 Virginia: 15
 North Carolina: 11
Present Distribution: Massachusetts (?) to West Indies and Texas.

FAMILY CAECIDAE

Caecum cooperi S. Smith, 1862

Caecum cooperi S. Smith, 1862, *Ann. Lyceum Nat. Hist. New York* 7: 154, 168.
Caecum cooperi Dall, 1892: 299.

Pleistocene Distribution:
 Virginia: 16
Present Distribution: South of Cape Cod to northern Florida.

Caecum pulchellum Stimpson, 1851

Plate 16, Figure 3

Caecum pulchellum Stimpson, 1851, *Shells of New England*, 36, pl. 2, fig. 3.

Pleistocene Distribution:
 Massachusetts: 1 (after Cushman)
Present Distribution: Cape Cod and south to North Carolina.

FAMILY VERMETIDAE

Vermetus nigricans Dall, 1883

Plate 16, Figure 4

Vermetus lumbricalis var. *nigricans* Dall, 1883, *Proc. U. S. Nat. Mus.* 6: 334.

Pleistocene Distribution:
 Virginia: 7
Present Distribution: Florida

FAMILY TRICHOTRÓPIDAE

Trichotropis borealis costellata Couthouy, 1838

Plate 16, Figure 5

Trichotropis borealis costellata Couthouy, 1838, *Boston Jour. Nat. Hist.* 2: 108.

Pleistocene Distribution:
 Hudson Bay: 4
 Labrador: 1, 2
 Quebec: 9
Present Distribution: Arctic to Massachusetts.

FAMILY TRIPHORIDAE

Triphora nigrocincta (Adams), 1839

Plate 17, Figure 1

Cerithium nigrocincta Adams, 1839, *Jour. Boston Soc. Nat. Hist.* 2: 286, pl. 4.
Triphora perversa var. *nigrocincta* Maury, 1922, *Bull. Amer. Paleont.* 9 (38): 91.

Sometimes considered a variety of the European *T. perversa* Linné.

Pleistocene Distribution:
 New Jersey: 16
 Maryland: 16
Present Distribution: Massachusetts to Florida.

FAMILY CERITHIOPSIDAE

Cerithiopsis greenii C. B. Adams, 1839

Plate 17, Figure 2

Cerithiopsis greenii C. B. Adams, 1839, *Jour. Boston Soc. Nat. Hist.* 2: 287, pl. 4, fig. 12.
Cerithiopsis greenii Dall, 1892: 269.

Pleistocene Distribution:
 Massachusetts: 1
Present Distribution: Cape Cod to both sides of Florida.

Cerithiopsis subulata (Montagu), 1808

Plate 17, Figures 4, 5

Murex subulatus Montagu, 1808, *Test. Brit. Suppl.*, 115, pl. 30, fig. 6.
Cerithiopsis subulata Dall, 1892: 268.

Pleistocene Distribution:
 New Jersey: 12, 16
 North Carolina: 1
 South Carolina: 12
Present Distribution: Massachusetts to West Indies.

Seila adamsii (Lea), 1845

Plate 17, Figure 3

Cerithium terebrale C. B. Adams, 1840, *Jour. Boston Soc. Nat. Hist.* 3: 320, pl. 3, fig. 7. Not of Lamarck.
Cerithium adamsii H. C. Lea, 1845, *Trans. Amer. Philos. Soc.* 9: 42.
Seila adamsii Dall, 1892: 267.

Closely related to the Miocene *Seila clavulus* from Virginia.

Pleistocene Distribution:
 New Jersey: 9, 12
 Maryland: 16
 Virginia: 7
 South Carolina: 14
Present Distribution: Massachusetts to Texas and West Indies.

FAMILY CERITHIIDAE

Bittium alternatum Say, 1822

Plate 17, Figure 6

Bittium alternatum Say, 1822, *Jour. Acad. Nat. Sci. Phila.* 2: 243.

Pleistocene Distribution:
Virginia : 13
Present Distribution: Gulf of St. Lawrence to Virginia.

Bittium cerithidioides Dall, 1889

Bittium cerithidioides Dall, 1889, *Bull. Mus. Comp. Zool.* **18**: 258.
Alabina cerithidioides Olssen *et al.*, 1953: 292, pl. 48, fig. 9.

Pleistocene Distribution:
South Carolina : 16
Present Distribution: North Carolina to West Indies.

FAMILY OVULIDAE

Simnia acicularis (Lamarck), 1811

Plate 17, Figures 11, 12

Ovula acicularis Lamarck, 1811, *Ann. Mus. Hist. Nat. Paris* **16**(92) : 112.
Volva acicularis Holmes, 1860: 79, pl. 12, fig. 13.

Pleistocene Distribution:
South Carolina : 14, 16
Present Distribution: North Carolina to West Indies.

FAMILY APORRHAIDAE

Aporrhais occidentalis Beck, 1836

Plate 20, Figures 7, 8

Aporrhais occidentalis Beck, 1836, *Magazine de Zool.* **6**, Classe 5, pl. 72 and text.

Pleistocene Distribution:
Labrador : (after Dawson and Packard) 2, 3
Present Distribution: Gulf of St. Lawrence to off North Carolina.

FAMILY STROMBIDAE

Strombus pugilis alatus Gmelin, 1791

Plate 17, Figure 8

Strombus pugilis Linné, 1758, *Syst. Nat.*, 10th ed., 744 (part).
Strombus alatus Gmelin, 1791, *Syst. Nat.*, 13th ed., 1 : 3513, no. 14.
Strombus pugilis Holmes, 1860: 61, pl. 10, figs. 1, 1a.
Strombus pugilis Dall, 1890: 177 (part).
Strombus pugilis alatus Olsson *et al.*, 1953: 273, pl. 34, figs. 7, 7a.

Pleistocene Distribution:
South Carolina : 16
Present Distribution: North Carolina to Florida; West Indies.

FAMILY CASSIDAE

Cassis madagascarensis Lamarck, 1822

Plate 17, Figure 9

Cassis madagascarensis Lamarck, 1822, *Animaux sans Vert.* **7**: 219.
Cassis madagascarensis Clench, 1944, *Johnsonia* **1**(16) : 14, pl. 7.

Pleistocene Distribution:
North Carolina : 26
Present Distribution: North Carolina to the West Indies and the Greater Antilles.

Phalium granulatum (Born), 1778

Plate 17, Figure 10

Buccinum granulatum Born, 1778, *Testacea Musei Caesari Vindobonensis*, 239.
Buccinum inflatum Shaw, 1811, *Nat. Misc.* 22, pl. 959.
Cassis sulcosa of authors. Not Bruguiere.
Phalium granulatum Clench, 1944, *Johnsonia* **1**(16) : 6, pl. 1, figs. 3–7; pl. 3, figs. 1–4.

Pleistocene Distribution:
North Carolina : 26
Present Distribution: North Carolina to the West Indies.

FAMILY CYMATIIDAE

Cymatium parthenopeum (von Salis), 1793

Plate 17, Figure 16

Murex costatum Born, 1778, *Index Museum Caesarei Vindbonensis*, 295. Not *Murex costatus* Pennant, 1777.
Murex parthenopeum von Salis, 1793, *Reisen in versch. Königreich Naepel* **1**: 370, pl. 7, fig. 4.
Cymatium (Monoplex) parthenopeum Clench and Turner, 1957, *Johnsonia* **3**(36) : 228; pl. 110, fig. 4; pl. 112, figs. 7, 8; pl. 113, figs. 9, 10; pl. 128, figs. 1–3.

Pleistocene Distribution:
North Carolina : 26
Present Distribution: North Carolina (?), Florida, and West Indies; world-wide.

FAMILY TONNIDAE

Ficus communis Röding, 1798

Plate 17, Figures 13, 14

Ficus communis Röding, 1798, *Museum Bolton*, 148.
Pyrula papyratia Say, 1822, *Jour. Acad. Nat. Sci. Phila.* **2**: 238.
Ficus papyratia Olsson *et al.*, 1953: 258, pl. 41, figs. 1–1b.

Pleistocene Distribution:
South Carolina : (after Pugh)
Present Distribution: North Carolina to Gulf of Mexico.

FAMILY CHORISTIDAE

Choristes elegans Carpenter, 1872

Plate 17, Figure 15

Choristes elegans Carpenter, 1872, in Dawson, Post Pliocene of Canada, *Canadian Naturl.* **6**, reprint, p. 88.

Pleistocene Distribution:
Quebec : 9
Present Distribution: Massachusetts (deep water).

FAMILY MURICIDAE

Murex (Muricanthus) fulvescens Sowerby, 1834

Plate 18, Figure 2

Murex fulvescens Sowerby, 1834, *Conch. Illust., Murex,* fig. 30, Catalogue, 7, sp. 94.
Murex spinicostatus Valenciennes, Holmes, 1860: 61, 10, fig. 2.
Murex fulvescens Clench and Farfante, 1945, *Johnsonia* 1(17): 42, pl. 22.

Pleistocene Distribution:
  Georgia: 2
Present Distribution: North Carolina to Texas.

Murex pomum Gmelin, 1791

Plate 18, Figure 1

Murex pomum Gmelin, 1791, *Syst. Nat.,* 13th ed., 1(6): 3527.
Murex pomum Dall, 1890: 142.
Murex pomum Clench and Farfante, 1945, *Johnsonia* 1: 26, pl. 14, figs. 1–3.
Murex pomum Olsson *et al.,* 1953: 243, pl. 34, fig. 1.

Pleistocene Distribution:
  South Carolina: 13, 14
Present Distribution: North Carolina to Gulf of Mexico and West Indies.

Eupleura caudata (Say), 1822

Plate 18, Figures 3, 4

Ranella caudata Say, 1822, *Jour. Acad. Nat. Sci. Phila.* 2: 236.
Eupleura caudata Holmes, 1860: 62, pl. 10, fig. 3.
Eupleura caudata Dall, 1890: 144.
Eupleura caudata Clark, 1906: 183, pl. 49, figs. 7, 8.
Eupleura caudata Olsson *et al.,* 1953: 256, pl. 37, figs. 6, 6a, 6b.

Pleistocene Distribution:
  Massachusetts: 1
  New York: 2
  New Jersey: 3, 7, 8, 9, 11, 12, 13, 17
  Maryland: 1, 2, 15, 16
  Virginia: 15
  North Carolina: 1, 9, 10, 21
  South Carolina: 7, 11, 14, 16
Present Distribution: Massachusetts to Florida.

Urosalpinx cinerea (Say), 1822

Plate 18, Figure 5

Fusus cinereus Say, 1822, *Jour. Acad. Nat. Sci. Phila.* 2: 236.
Fusus cinereus Holmes, 1860: 68, pl. 11, fig. 5.
Urosalpinx cinereus Clark, 1906: 184, pl. 49, figs. 9, 10.

Pleistocene Distribution:
  Massachusetts: 1
  New York: 2
  New Jersey: 3, 7, 12, 13, 16, 17
  Maryland: 16
  Virginia: 16
  North Carolina: 1, 11, 21
  South Carolina: 11, 12, 14, 16
Present Distribution: Nova Scotia to southern Florida. Prince Edward Island to Florida.

Trophon clathratus (Linné), 1767

Plate 18, Figures 6, 7

Murex clathratus Linné, 1767, *Syst. Nat.,* 12th ed. 1: 1223.
Trophon clathratus Feyling-Hanssen, 1955: 163.

Pleistocene Distribution:
  Quebec: 4, 5, 9
  Massachusetts: 1
Present Distribution: Nova Scotia to Hatteras; deep water.

Trophon scalariforme Gould, 1840

Trophon scalariformis Gould, 1840, *Amer. Jour. Sci.* 38: 197.

Pleistocene Distribution:
  Quebec: 4, 5, 9
Present Distribution: Labrador to Massachusetts.

This may be a variety of *T. clathratus* L.

Thais lapillus (Linné), 1758

Plate 18, Figures 8, 9

Buccinum lapillus Linné, 1758, *Syst. Nat.,* 10th ed., 739.
Purpura lapillus Lamarck, 1803, *Ann. Mus. Hist. Nat. Paris* 2: 64.
Thais (Polytropa) lapillus Clench, 1947, *Johnsonia* 2 (23): 86.

Pleistocene Distribution:
  Maine: 18
  New York: 3
Present Distribution: Newfoundland to Connecticut.

Thais haemastoma floridana (Conrad), 1837

Plate 18, Figures 10–12

Purpura floridana Conrad, 1837, *Jour. Acad Nat. Sci. Phila.* 7: 265, pl. 20, fig. 21.
Thais haemastoma floridana Clench, 1947, *Johnsonia* 2 (23): 76, pl. 37, figs. 1–4.

Pleistocene Distribution:
  New Jersey: 12, 13, 16, 17
  Maryland: 1
  Virginia: 11
Present Distribution: North Carolina to Texas and Caribbean.

FAMILY COLUMBELLIDAE

Columbella (Anachis) obesa (Adams), 1845

Plate 18, Figures 14, 15

Buccinum obesum Adams, 1845, *Proc. Boston Soc. Nat. Hist.* 2: 2.
Columbella ornata Holmes, 1860: 74, pl. 12, figs. 6, 6a.

Pleistocene Distribution:
  North Carolina: 10, 11
  South Carolina: 12, 14, 16
Present Distribution: North Carolina to Florida and West Indies.

Columbella (Anachis) avara Say, 1822

Plate 18, Figure 16

Columbella avara Say, 1822, *Jour. Acad. Nat. Sci. Phila.* 2: 230.
Columbella avara Holmes, 1860: 73, pl. 12, fig. 4.
Anachis avara Dall, 1890: 135.

Pleistocene Distribution:
New Jersey: 9, 12, 13, 16, 17
Maryland: 1
Virginia: 16
North Carolina: 1, 9, 10
South Carolina: 2, 14, 16
Present Distribution: Massachusetts to Florida.

Columbella (Astyris) lunata (Say), 1826

Plate 18, Figures 22, 23

Nassa lunata Say, 1826, *Jour. Acad. Nat. Sci. Phila.* 5: 213.
Columbella lunata Holmes, 1860: 74, pl. 12, figs. 5. 5a.
Astyris lunata Dall, 1890: 137.
Columbella (Astyris) lunata Clark, 1906: 182, pl. 49, figs. 5, 6.

Pleistocene Distribution:
Massachusetts: 1
New York: 3
New Jersey: 17
Maryland: 2, 15, 16
Virginia: 11
North Carolina: 1, 9, 10, 11, 18
South Carolina: 14, 16
Present Distribution: Prince Edward Island to the Gulf of Mexico.

Family NASSARIIDAE

Nassarius obsoletus (Say), 1822

Plate 18, Figures 18, 20, 21

Nassa obsoleta Say, 1822, *Jour. Acad. Nat. Sci. Phila.* 2: 232.
Buccinum obsoletum Holmes, 1860: 71, pl. 12, fig. 1.
Ilyanassa obsoleta Clark, 1906: 181, pl. 49, figs. 3, 4.

Pleistocene Distribution:
Massachusetts: 1, 3
New York: 2, 7
New Jersey: 3, 5, 7, 8, 9, 11, 12, 13, 16, 17
Delaware: 1, 2, 3
Maryland: 1, 2, 15, 16
Virginia: 11, 13
North Carolina: 1, 3, 9, 10, 11, 13, 21
South Carolina: 1, 2, 9, 11, 12, 14, 16
Present Distribution: Gulf of St. Lawrence to Florida.

Nassarius trivittatus (Say), 1822

Plate 18, Figure 19

Nassa trivittata Say, 1822, *Jour. Acad. Nat. Sci. Phila.* 2: 231.
Buccinum trivittatum Holmes, 1860: 72, pl. 12, fig. 2.
Nassa trivittata Clark, 1906: 180, pl. 49, figs. 1, 2.

Pleistocene Distribution:
Massachusetts: 1, 3

New York: 2, 3, 5, 6, 7
New Jersey: 3, 5, 7, 8, 9, 11, 12, 13, 16, 17, 25
Delaware: 3
Maryland: 1, 15, 16
Virginia: 11, 12, 15
North Carolina: 1, 3, 5, 6, 7, 8, 9, 10, 11, 13, 18, 21, 24
South Carolina: 7, 16
Present Distribution: Gulf of St. Lawrence to Florida.

Nassarius vibex (Say), 1822

Plate 18, Figure 13

Nassa vibex Say, 1822, *Jour. Acad. Nat. Sci. Phila.* 2: 231.
Nassa vibex Dall, 1890: 132.
Nassarius (Phrontis) vibex Olsson *et al.*, 1953: 220, pl. 33, figs. 1, 1a.

Pleistocene Distribution:
New York: 3
New Jersey: 3, 7, 8, 9, 11, 12, 13, 16, 17
Delaware: 3
Maryland: 1
Virginia: 11
North Carolina: 1
South Carolina: 11, 12
Present Distribution: Massachusetts (?) to the Gulf of Mexico and the West Indies.

Nassarius acutus (Say), 1822

Plate 18, Figure 17

Nassa acuta Say, 1822, *Jour. Acad. Nat. Sci. Phila.* 2: 234.
Nassa acuta Say, 1834, *Amer. Conch.*, 254, pl. 57, fig. 3.
Buccinum acutum Holmes, 1860: 72, pl. 12, fig. 3.

Pleistocene Distribution:
New Jersey: 17
Virginia: 15
North Carolina: 1, 6, 7, 9, 10, 11, 13
South Carolina: 12, 14, 16
Georgia: 1
Present Distribution: North Carolina to Texas.

Family BUCCINIDAE

Buccinum undatum Linné, 1758

Plate 18, Figure 24

Buccinum undatum Linné, 1758, *Syst. Nat.*, 10th ed., 740.
Buccinum undatum Feyling-Hanssen, 1955: 166.

Pleistocene Distribution:
Labrador: 1
Newfoundland: 3, 7, 8, 26
Quebec: 4
New Brunswick: 3, 8
Maine: 3, 8, 19
Massachusetts: 1
New Jersey: 16, 17
Present Distribution: Arctic to Maryland.

Buccinum cyaneum Bruguiere, 1789

Plate 18, Figure 29; Plate 19, Figures 1, 3

Buccinum cyaneum Bruguiere, 1789, *Encycl. Méthod., Hist. Nat des Verms* 1: 266.
Buccinum groenlandicum Mörch, 1857.

Pleistocene Distribution:
  Labrador: 6
  Quebec: 4
  Maine: 2, 15, 18
Present Distribution: Arctic to Cape Cod.

Buccinum tenue Gray, 1839

Plate 19, Figures 5, 6

Buccinum tenue Gray, 1839, *Beechey's Voyages, Zoology,* 128, pl. 36, fig. 19.

Pleistocene Distribution:
  Hudson Bay: 1, 6
  Newfoundland: 1, 9, 16, 34
  James Bay: 1, 2, 3
  Labrador: 2
  Maine: 6, 13, 19
Present Distribution: Arctic to Maine.

Buccinum plectrum Stimpson, 1865

Plate 18, Figures 28, 30

Buccinum plectrum Stimpson, 1865, *Canadian Natur.,* n.s., **2**: 374.

Pleistocene Distribution:
  Quebec: 4
  Maine: 2
Present Distribution: Arctic to the Gulf of St. Lawrence (including variety *carinatum*).

Buccinum ciliatum Fabricius, 1780

Plate 18, Figure 25

Tritonium ciliatum Fabricius, 1780, *Fauna Groenl.,* 401.
Buccinum ciliatum Feyling-Hanssen, 1955: 168.

Pleistocene Distribution:
  Quebec: 4, 9
Present Distribution: Greenland and Gulf of St. Lawrence; 3–112 fathoms.

Buccinum tottenii Stimpson, 1865

Plate 19, Figure 2

Buccinum tottenii Stimpson, 1865, *Canadian Natur., n.s.,* **2**: 385.
Buccinum terrae-novae Mörch.
Buccinum totteni Feyling-Hanssen, 1955: 169.

Pleistocene Distribution:
  Quebec: 4
  Maine: 2
Present Distribution: Labrador to Gulf of St. Lawrence.

Buccinum kroyeri (Möller), 1842

Plate 19, Figures 7, 8

Fusus kroyeri Möller, 1842, *Naturhist. Tidsskrift* **4**: p. 88; *Index. Moll. Groenl.,* 15.

Pleistocene Distribution:
  Labrador
  Quebec: 4
Present Distribution: Circumboreal; Greenland to Gulf of St. Lawrence.

Buccinum scalariformis Beck, 1842

Plate 19, Figure 4

Buccinum scalariforme Beck, 1842, *Naturhist. Tidsskrift* **4**: 84.

Pleistocene Distribution:
  Quebec: 4
Present Distribution: Circumboreal.

Buccinum glaciale Linné, 1761

Plate 19, Figure 9

Buccinum glaciale Linné, 1761, *Fauna Sueicia,* 2nd ed., 523.
Buccinum donovanni Gray, 1839, *Beechey's Voyages, Zoology,* 128.

Pleistocene Distribution:
  Quebec: 4
  Maine: 2
Present Distribution: Greenland to Gulf of St. Lawrence.

Buccinum tumidulum Sars, 1878

Plate 18, Figures 26, 27

Buccinum tumidulum Sars, 1878, *Moll. Reg. Arct. Norv.,* 263.

Pleistocene Distribution:
  Reed Island, N.W.T.
Present Distribution: Greenland; Arctic.

Cantharus cancellaria (Conrad), 1846

Plate 19, Figures 19, 20

Pollia cancellaria Conrad, 1846, *Proc. Acad. Nat. Sci. Phila.* **3**: 25.

Pleistocene Distribution:
  New Jersey: 12
  North Carolina: 1
  South Carolina: 11, 12, 14, 16
Present Distribution: South Carolina to Yucatan.

Cantharus tinctus (Conrad), 1846

Plate 19, Figures 17, 18

Pollia tinctus Conrad, 1846, *Proc. Acad. Nat. Sci. Phila.* **3**: 25.

Pleistocene Distribution:
  South Carolina: 16
Present Distribution: North Carolina to both sides of Florida and the West Indies.

FAMILY **NEPTUNEIDAE**

Colus stimpsoni stimpsoni (Mörch), 1867

Plate 19, Figures 11, 12

Fusus stimpsoni Mörch, 1867, *Vid. Medd. Naturh. Foren.* (Copenhagen), 83.
Fusus islandicus Gould, 1841. Not Gmelin.

Pleistocene Distribution:
Maine: 10
Massachusetts: 1
Present Distribution: Arctic to Labrador.

Colus stimpsoni liratulus (Verrill), 1882

Plate 19, Figures 10, 14

Sipho stimpsoni var. *liratulus* Verrill, 1882, *Trans. Conn. Acad.* **5**: 500.

Pleistocene Distribution:
New Jersey: 17
Present Distribution: Maine to Massachusetts; 10 to 17 fathoms.

Neptunea decemcostata (Say), 1826

Plate 20, Figure 2

Fusus decemcostatus Say, 1826, *Jour. Acad. Nat. Sci. Phila.* **5**: 214.

Pleistocene Distribution:
Maine: 2
Massachusetts: 1, 3
New York: 3
New Jersey: 25; beach wash
Present Distribution: Nova Scotia to Massachusetts.

Neptunea despecta tornata (Gould), 1839

Plate 20, Figures 3, 10; Plate 21, Figure 21

Murex despecta Linné, 1758, *Syst. Nat.*, 10th ed., 754; 12th ed., 1222 (1767) (part).
Fusus tornata Gould, 1839, *Amer. Jour. Sci.* **38**: 197.

Pleistocene Distribution:
James Bay: 2
Newfoundland: 3
Labrador: 2, 3
Quebec: 4, 5, 7, 9
Maine: 2, 4, 15, 18
Present Distribution: Gulf of St. Lawrence to off Massachusetts; 10 to 471 fathoms.

Neptunea stonei (Pilsbry), 1893

Plate 19, Figure 13

Chrysodomus stonei Pilsbry, 1893, *Nautilus* **7**: 67-69.

Pleistocene Distribution:
Massachusetts: 1 (?)
New York: 3, 5

New Jersey: 17, 25; beach wash
North Carolina: 24, 26
Present Distribution: Extinct.

Busycon canaliculatum (Linné), 1758

Plate 19, Figures 16, 21

Murex canaliculatus Linné, 1758, *Syst. Nat.*, 10th ed., 753.
Busycon canaliculatum Holmes, 1860: 66, pl. 11, fig. 3.
Busycon canaliculatum Clark, 1906: 180, pls. 46, 47, 48.

Pleistocene Distribution:
New York: 3
New Jersey: 3, 5, 8, 9, 11, 12, 13, 16, 17
Delaware: 3
Maryland: 15
Virginia: 11, 16
North Carolina: 1, 5, 7, 8, 9, 10, 13
South Carolina: 11, 12, 16
Present Distribution: Massachusetts to Florida.

Busycon carica (Gmelin), 1790

Plate 20, Figure 1

Murex carica Gmelin, 1790, *Syst. Nat.*, 13th ed., 3545.
Busycon carica Holmes, 1860: 65, pl. 11, fig. 1.
Fulgur carica Clark, 1906: 179, pls. 43, 44, 45.

Pleistocene Distribution:
New Jersey: 3, 8, 9, 11, 12, 13, 17
Maryland: 1, 15, 16
Virginia: 11, 15
North Carolina: 1, 3, 8, 9, 10, 15
South Carolina: 2, 7, 14, 16
Present Distribution: Massachusetts to Florida.

Busycon perversum (Linné), 1758

Plate 19, Figures 15, 22

Murex perversus Linné, 1758, *Syst. Nat.*, 10th ed., 753.
Busycon perversum Holmes, 1860: 65, pl. 11, fig. 2.
Busycon perversum Dall, 1890: 116.

Pleistocene Distribution:
New Jersey: 13, 16
Delaware: 3
Virginia: 11
North Carolina: 1
South Carolina: 2, 14, 16
Present Distribution: North Carolina to Texas.

According to Abbott (1954: 236), most of the specimens referred to *B. perversum* from the Atlantic coast north of Florida should preferably be identified as *B. contrarium* Conrad.

Busycon spiratum (Lamarck), 1816

Plate 20, Figure 4

Pyrula spiratum Lamarck, 1816, *Ency. Méthod.* (*Vers*), pt. 433: 8.
Bulla pyrum Dillwyn, 1817, *Des. Cat. Rec. Shells* **1**: 485.

Pyrula pyrum Holmes, 1860: 67, pl. 11, fig. 4.
Fulgur pyrum Dall, 1890: 112.

Pleistocene Distribution:
   South Carolina: 16
Present Distribution: North Carolina to Florida and Texas.

FAMILY **FASCIOLARIIDAE**

Fasciolaria distans Lamarck, 1822

Plate 20, Figure 6

? *Pyrula hunteria* Perry, 1811, *Conchology*, app., pl. 50.
Fasciolaria distans Lamarck, 1822, *Animaux sans Vert.* 7: 119.
Fasciolaria distans Holmes, 1860: 63, pl. 10, fig. 5.
Fasciolaria distans Dall, 1890: 102, pl. 7, fig. 10.

Pleistocene Distribution:
   South Carolina: 14, 16
Present Distribution: North Carolina to Florida, Texas.

If *Pyrula hunteria* Perry is synonymous with *Fasciolaria distans* Lamarck, as has been suggested, this name should be used instead of the better known name *F. distans.*

Fasciolaria gigantea Kiener, 1840

Plate 20, Figure 11

Fasciolaria gigantea Kiener, 1840, *Spec. Coquilles (Fasciolaria)*, 5, pls. 10, 11.
Fasciolaria gigantea Holmes, 1860: 63, pl. 10, fig. 4.
Fasciolaria gigantea Olsson *et al.*, 1953: 216.

Pleistocene Distribution:
   South Carolina: 14, 16
Present Distribution: North Carolina to both sides of Florida.

Fasciolaria ligata Mighels and Adams, 1842

Plate 20, Figure 9

Fasciolaria ligata Mighels and Adams, 1842, *Boston Jour. Nat. Hist.* 4: 51.
Ptychatractus ligatus Johnson, 1934: 127.

Pleistocene Distribution:
   Quebec: 9 (very rare, after Dawson)
Present Distribution: Gulf of St. Lawrence to Connecticut; 15–60 fathoms.

Fasciolaria tulipa (Linné), 1758

Plate 20, Figure 5

Murex tulipa Linné, 1758, *Syst. Nat.*, 10th ed., 754.
Fasciolaria tulipa Dall, 1890: 101, pl. 7, fig. 11.

Pleistocene Distribution:
   South Carolina: 16
Present Distribution: North Carolina to West Indies and Texas.

"Fusus" minor Holmes, 1860

Plate 21, Figure 25

Fusus minor Holmes, 1860: 68, pl. 11, figs. 6, 6a.

Pleistocene Distribution:
   South Carolina: 16 (after Holmes)
Present Distribution: Unknown.

"Fusus" rudis Holmes, 1860

Plate 21, Figure 26

Fusus. rudis Holmes, 1860: 70, pl. 11, figs. 11, 11a.

Pleistocene Distribution:
   South Carolina: 16 (after Holmes)
Present Distribution: Unknown.

"Fusus" bullata Holmes, 1860

Plate 21, Figure 27

Fusus bullata Holmes, 1860: 69, pl. 11, figs. 10, 10a. Not *Fusinus bullatus* Dall, 1927.

Pleistocene Distribution:
   South Carolina: 14 (after Holmes)
Present Distribution: Unknown.

"Fusus" conus Holmes, 1860

Plate 21, Figures 24, 28

Fusus conus Holmes, 1860: 69, pl. 11, figs. 7, 8.

Pleistocene Distribution:
   South Carolina: 14 (after Holmes)
Present Distribution: Unknown.

"Fusus" filiformis Holmes, 1860

Plate 21, Figure 29

Fusus filiformis Holmes, 1860: 69, pl. 11, fig. 9.

Pleistocene Distribution:
   South Carolina: 16 (after Holmes)
Present Distribution: Unknown.

FAMILY **MITRIDAE**

Mitra wandoensis (Holmes), 1860

Plate 21, Figure 23

Volutomitra wandoensis Holmes, 1860: 77, pl. 17, figs. 10, 10a.

Pleistocene Distribution:
   South Carolina (after Holmes).   May be pre-Pleistocene.
Present Distribution: North Carolina to Gulf of Mexico; 12–440 fathoms.

Prunum roscidum (Redfield), 1860

Plate 21, Figure 10

Marginella roscidum Redfield, 1860, *Proc. Acad. Nat. Sci. Phila.*
   12: 174.
Prunum roscidum Abbott, 1957, *Nautilus* 71: 52.

Pleistocene Distribution:
   Virginia: 15
   North Carolina: 1
   South Carolina: 13, 14, 16
Present Distribution: New Jersey to Florida.

This species has been confused in the literature with
Marginella apicina Menke, *M. guttata* Dillwyn, and *M.
limatula* Conrad.

FAMILY **OLIVIDAE**

Oliva sayana Ravenel, 1834

Plate 21, Figure 1

Oliva litterata Lamarck, 1822, *Animaux sans Vert.*, 425. Not
   Porphyria litterata Bolton.
Oliva sayana Ravenel, 1834, *Catalogue*, 19.
Strephona litterata Holmes, 1860: 75, pl. 12, fig. 7.
Oliva litterata Dall, 1890: 44.

Pleistocene Distribution:
   North Carolina: 1, 4, 6, 7, 8, 9, 13
   South Carolina: 1, 7, 8, 12, 14, 16
Present Distribution: North Carolina to Texas.

Olivella mutica (Say), 1822

Plate 21, Figures 11, 12

Oliva mutica Say, 1822, *Jour. Acad. Nat. Sci. Phila.* 2: 228.
Oliva mutica Holmes, 1860: 76, pl. 12, fig. 8.
Olivella mutica Dall, 1890: 45.
Olivella (Dactylidia) mutica Olsson *et al.*, 1953: 186, pl. 39,
   fig. 7.

Pleistocene Distribution:
   New Jersey: 17
   Virginia: 12, 15
   North Carolina: 1, 5, 6, 8, 9, 11
   South Carolina: 7, 9, 14, 16
Present Distribution: North Carolina to Texas and
West Indies.

FAMILY **TEREBRIDAE**

Terebra concava (Say), 1827

Plate 21, Figure 8

Turritella concava Say, 1827, *Jour. Acad. Nat. Sci. Phila.*
   1(5): 207.
Terebra concava Dall, 1890: 24.
Terebra (Strioterebrum) concava Olsson *et al.*, 1953: 168, pl.
   58, figs. 9, 9a.
Terebra (Strioterebrum) concava Gardner, 1948, *U. S. Geol.
   Surv. Prof. Paper* 199-B: 277, pl. 38, fig. 32.

Pleistocene Distribution:
   New Jersey: 9, 11, 12, 13, 16
   North Carolina: 1, 7, 10, 11
   South Carolina: 16
Present Distribution: North Carolina to Florida.

Terebra dislocata (Say), 1822

Plate 21, Figures 2, 3

Cerithium dislocatum Say, 1822, *Jour. Acad. Nat. Sci. Phila.*
   2: 235.
Terebra dislocata Holmes, 1860: 70, pl. 11, fig. 12.
Terebra dislocata Dall, 1890: 24.
Terebra dislocata Clark, 1906: 177, pl. 42, figs. 7, 8.
Terebra (Strioterebra) dislocata Olsson *et al.*, 1953: 167.

Pleistocene Distribution:
   New Jersey: 11, 12, 13, 16
   Maryland: 16
   North Carolina 1, 6, 7, 8, 9, 10, 11
   South Carolina: 2, 7, 11, 14, 16
Present Distribution: Virginia to Florida to Texas.

Terebra protexta (Conrad), 1846

Plate 21, Figure 9

Cerithium protextum Conrad, 1846, *Proc. Acad. Nat. Sci. Phila.*
   3: 26.
Terebra (Acus) protexta Dall, 1890: 25.
Terebra (Strioterebrum) protexta Gardner, *U. S. Geol. Surv.
   Prof. Paper* 199-B: 277, pl. 38, fig. 31.
Terebra (Strioterebrum) protexta Olsson *et al.*, 1953: 168.

Pleistocene Distribution:
   South Carolina: 16
Present Distribution: North Carolina to Florida and
Texas.

FAMILY **TURRIDAE**

Lora Gistel, 1848 (= Bela Gray, 1847)

The species of this genus were not studied in the
present investigation. The following records are taken
from the literature:

Lora elegans (Möller): Montreal.
Lora pleurotomaria Couthouy (= *L. pyramidalis* of
   authors): Montreal.
Lora scalaris Möller (= *L. turricula* of authors):
   Montreal; Rivière du Loup; Hudson Bay.
Lora treveliana Turton: Rivière du Loup; Labrador.
Lora violacea Mighels and Adams: Montreal.
Lora cancellata Mighels and Adams: Murray Bay;
   Labrador; Maine.
Lora harpularia Couthouy: Montreal; Quebec; Rivière
   du Loup; Murray Bay.
Lora incisula Verrill: James Bay.
Lora americana Packard: James Bay.

Mangilia cerina (Kurtz and Stimpson), 1851

Plate 21, Figure 6

Pleurotoma cerinum Kurtz and Stimpson, 1851, *Proc. Boston Soc. Nat. Hist.* **4**: 11.
Pleurotoma cerinum Holmes, 1860: 77, pl. 12, figs. 9, 9a.
Mangilia cerina Clark, 1906: 178, pl. 42, figs. 9, 10.

Pleistocene Distribution:
New Jersey: 16
Maryland: 2, 15, 16
North Carolina: 1, 10, 11
South Carolina: 14, 16
Present Distribution: Massachusetts to Florida.

Mangilia plicosa (Adams), 1850

Plate 21, Figure 7

Pleurotoma plicata C. B. Adams, 1840, *Boston Jour. Nat. Hist.* **3**: 318, pl. 3, fig. 6. Not of Lamarck.
Pleurotoma plicosa Adams, 1850, *Contrib. Conchol.*, 54.
Mangilia plicosa Dall, 1890: 41.

Pleistocene Distribution:
New Jersey: 12, 16
Present Distribution: Massachusetts to Florida.

Mangilia stellata Stearns, 1872

Plate 21, Figures 4, 5

Mangilia stellata Stearns, 1872, *Proc. Boston Soc. Nat. Hist.* **15**: 22.
Mangilia stellata Dall, 1890: 41.

Pleistocene Distribution:
New Jersey: 16
Present Distribution: Florida.

FAMILY **CANCELLARIIDAE**

Cancellaria reticulata (Linné), 1767

Plate 21, Figure 14

Voluta reticulata Linné, 1767, *Syst. Nat.*, 12th ed., 1190.
Cancellaria reticulata Holmes, 1860: 64, pl. 10, fig. 6.

Pleistocene Distribution:
South Carolina: 16
Present Distribution: North Carolina to Gulf of Mexico.

Cancellaria venusta Tuomey and Holmes, 1856

Plate 21, Figure 22

Cancellaria venusta Tuomey and Holmes, 1856, *Pleiocene Fossils of South Carolina*, 144, pl. 23, fig. 18.
Cancellaria venusta Holmes, 1860: 64, pl. 10, fig. 7.

Pleistocene Distribution:
South Carolina (after Holmes)
Present Distribution: Unknown; may be variety of *C. reticulata.*

FAMILY **ACTEOCINIDAE**

Retusa canaliculata (Say), 1826

Plate 21, Figure 13

Volvaria canaliculata Say, 1826, *Jour. Acad. Nat. Sci. Phila.* **5**: 211.
Volvaria canaliculata Holmes, 1860: 78, pl. 12, figs. 11, 11a.
Acteocina canaliculata Olsson *et al.*, 1953: 159, pl. 25, figs. 6, 6a, 6b.

Pleistocene Distribution:
New York: 3
New Jersey: 3, 4, 16, 17
Maryland: 15, 16
Virginia: 15
North Carolina: 1, 6, 7, 9, 10, 11, 18
South Carolina: 2, 14, 16
Georgia: 1
Present Distribution: Gulf of St. Lawrence to Texas and West Indies.

Retusa pertenuis (Mighels), 1842

Plate 16, Figure 11

Bulla pertenuis Mighels, 1842, *Boston Jour. Nat. Hist.* **4**: 346.

Pleistocene Distribution:
Quebec: 9 (after Dawson)
Present Distribution: Greenland to Florida; 10 to 294 fathoms.

FAMILY **SCAPHANDRIDAE**

Diaphana debilis (Gould), 1840

Plate 17, Figure 7

Bulla debilis Gould, 1840, *Amer. Jour. Sci.*, 1st ser., **38**: 196.

Pleistocene Distribution:
Quebec: 9 (after Dawson)
Present Distribution: Greenland to Connecticut.

Cylichna alba Brown, 1827

Plate 21, Figure 18

Cylichna alba Brown, 1827, *Illustr. Conchol. of Great Britain*, 3, pl. 19, figs. 43, 44.

Pleistocene Distribution:
James Bay: 1, 2, 3
Newfoundland: 16
Vermont: 6
Quebec: 4, 9
Present Distribution: Arctic to North Carolina.

Cylichna oryza Totten, 1835

Plate 16, Figure 10

Cylichna oryza Totten, 1835, *Amer. Jour. Sci.* **28**: 350.

Pleistocene Distribution:
Quebec: 9 (after Dawson)
Present Distribution: Maine to Connecticut.

Cylichna occulata Mighels, 1841

Plate 16, Figure 7

Cylichna occulata Mighels, 1841, *Proc. Boston Soc. Nat. Hist.* 1: 50.

Pleistocene Distribution:
Quebec: 5, 9
Maine
Present Distribution: Greenland to Maine.

FAMILY **AKERIDAE**

Haminoea solitaria (Say), 1822

Plate 21, Figures 16, 17

Bulla solitaria Say, 1822, *Jour. Acad. Nat. Sci. Phila.* 2: 245.

Pleistocene Distribution:
Quebec: 9 (after Dawson)
Maryland: 15
Present Distribution: Massachusetts to Georgia.

FAMILY **ELLOBIIDAE**

Melampus lineatus Say, 1822

Plate 21, Figures 19, 20

Melampus lineatus Say, 1822, *Jour. Acad. Nat. Sci. Phila.* 2: 246.
Melampus bidentatus Say, Holmes, 1860: 97, pl. 14, figs. 15, 15a. Not *M. bidentatus* Montagu.

Pleistocene Distribution:
New Jersey: 7, 12, 15, 16
Maryland: 1
South Carolina: 12
Present Distribution: Prince Edward Island to Gulf of Mexico.

XI. ADDITIONAL RECORDS

The species listed below have been reported from Pleistocene deposits, but have not been found by recent collectors, nor have they been verified by a study of material in various museums. Some may be misidentifications, for example the use of European specific names for American specimens, as has been done by Dawson and others. A few species, which have been obviously reworked from the underlying Pliocene, have been omitted from the list:

PELECYPODA

Macoma inflata Stimpson. Montreal; Rivière du Loup. Dawson.
Astarte arctica Gray. Murray Bay. Dawson.

Yoldia truncata Brown = *Yoldia glacialis* Wood.
Arca limula Conrad. Nantucket, Mass. Cushman.
Arca plicatura Conrad. Heislerville, N. J. U. S. Nat. Mus.
Pandora crassidens Conrad. Nantucket, Mass. Wilson; Cushman.
Macoma incongrua [4] Von Martens. Nantucket, Mass. Wilson; Cushman.
Serripes laperousii [4] Deshayes. Nantucket, Mass. Wilson; Cushman.

GASTROPODA

Acrica eschrichti Holboll (= *A. costulata* M. & A.?). Montreal; Quebec; Rivière du Loup. Dawson.
Trichotropis arctica Middendorff. Montreal. Dawson.
Admete viridula Fabricius. James Bay; Montreal. Dawson.
Philine lineolata Couthouy. Montreal. Dawson.
Cylichna nucleola Reeve. Montreal ? Dawson.
Cylichna striata Brown. Rivière du Loup. Dawson.
Bulla pertenuis Mighels. Montreal. Dawson.
Capulus commodus Middendorff. Levis. Dawson.
Cyclostrema costulata Möller. Montreal. Dawson.
Cyclostrema cutleriana Clark. Montreal. Dawson.
Rissoa exarata Stimpson. Montreal. Dawson.
Rissoa scobiculata Möller. Montreal. Dawson.
Skenea planorbis F. and H. Nantucket. Cushman.
Scala fragilis Gray ? Nantucket. Cushman.
Scalaria cf. *semanae* Orbigny. Maryland. E. R. Smith.
Rissoa aculeus Gould. Nantucket. Cushman.
Columbella (Astyris) holbolli Möller. Rivière du Loup. Dawson.
Chrysodomus spitzbergensis Reeve. Montreal. Dawson.
Margarites umbilicus B. & S. Hudson Bay. Richards.
Sipho ventricosus (Gray). James Bay. Richards, 1936.
Serpulorbis decussta Gmelin. South Carolina. Pugh.
Pyramidella sp. Maryland. Blake.

SCAPHOPODA

Dentalium cf. *entale simpsoni* Henderson. Maryland. Blake.

AMPHINEURA

Chaetopleura apiculata (Say). South Carolina. Holmes.

[4] *Macoma incongrua* and *Serripes leperousii* are Pacific coast species, not otherwise known east of Point Barrow, Alaska. According to Wilson (1905), their presence in the Pleistocene of Nantucket suggests a more open sea connection to the Pacific via a "northwest passage" during an interglacial stage.

BIBLIOGRAPHY

ABBOTT, R. TUCKER. 1954. American sea shells. New York, Van Nostrand.

ADDICOTT, WARREN O., and WILLIAM K. EMERSON. 1959. Late Pleistocene invertebrates from Punta Cabras, Baja California, Mexico. *Amer. Mus. Nat. Hist. Novitates* No. 1925 : 1–33.

ADIE, RAYMOND J. 1953. New evidence of sea-level changes in the Falkland Islands. *Falkland Islands Dependencies Survey Scientific Rept.* No. 9.

AKERS, W. H., and J. J. HOLCK. 1957. Pleistocene beds near the edge of the continental shelf, southeastern Louisiana. *Bull. Geol. Soc. Amer.* 68 : 985–992.

ALTSCHULER, Z. S., and E. J. YOUNG. 1960. Residual origin of the "Pleistocene" sand mantle in central Florida Uplands and its bearing on marine terraces and Cenozoic uplift. *U. S. Geol. Surv. Prof. Paper* 400-B : 202–207.

ANDERSON, C. A. 1950. E. W. Scripps 1940 cruise to the Gulf of California: Part 1: Geology of islands and neighboring land areas. *Geol. Soc. Amer. Memoir* 71 : 1–378.

ANDERSON, F. M. 1927. Nonmarine Tertiary deposits of Colombia. *Bull. Geol. Soc. Amer.* 38 : 591–644.

ANTEVS, ERNST. 1928. Shell beds on the Skagerack. *Geol. Fören.* 50 : 479–750. Stockholm.

—— 1929. Quaternary marine terraces in non-glaciated regions and changes of level of sea and land. *Amer. Jour. Sci.*, Ser. 5, 17 : 35–49.

——1939. Late Quaternary upwarpings of northeastern North America. *Jour. Geol.* 47 : 707–720.

ARMSTRONG, J. E., and W. L. BROWN. 1954. Late Wisconsin drift and associated sediments of the lower Fraser Valley, British Columbia, Canada. *Bull. Geol. Soc. Amer.* 65 : 349–364.

ARNOLD, RALPH. 1903. Paleontology and stratigraphy of the marine Pliocene and Pleistocene of San Pedro, California. *Memoir Calif. Acad. Sci.* 3.

ASKELSSON, JOHANNES. 1960a. Fossiliferous xenoliths in the Moberg formation of south Iceland. *Acta Naturalia Islandica* 2 (3).

—— 1960b. Pliocene and Pleistocene fossiliferous deposits. *In* On the geology and geophysics of Iceland, 21st Internat. Geol. Cong, 28–32. Reykjavik.

AUER, VÄINÖ. 1959. The Pleistocene of Fuego-Patagonia. Part III : Shoreline displacements. *Annales Acad. Sci. Fennicae*, Ser. A, No. 60. Helsinki.

BADEN-POWELL, D. F. W. 1948. The Pliocene-Pleistocene boundary in the British deposits. 18th Internat. Geol. Cong. London, Pt. 9 : 8–10.

BALDWIN, E. M. 1945. Some revisions of the late Cenozoic stratigraphy of the southern Oregon coast. *Jour. Geol.* 53 : 35–46.

BARDASON, G. S. 1925. A stratigraphic survey of the Pliocene deposits at Tjörnes in northern Iceland. *Der Kgl. Danske Vidensk. Selsk. Biol. Medd.* IV. 5. Köbenhavn.

BARTON, D. C. 1930. Surface geology of coastal southeast Texas. *Bull. Amer. Assoc. Petrol. Geol.* 14 : 1301–1320.

BEAL, C. H. 1948. Reconnaissance of the geology and the oil possibility of Baja California. *Geol. Soc. Amer. Memoir* 31.

BELL, ROBERT. 1890. Glacial phenomena in Canada. *Bull. Geol. Soc. Amer.* 1 : 287–310.

BELOV, N. A., and N. N. LAPINA. 1958. Nov'e dann'e o stratifikazii donn'x otlojenii Arkticheskogo Basseina severnogo Ledovitogo Okeana. (New data on the sequence of bottom sediments in the Arctic Ocean Basin.) *Dokladi Akad. Nauk, U.S.S.R.* 122 (11) : 115–118.

BERRY, E. W., and A. C. HAWKINS. 1935. Flora of the Pensauken formation. *Bull. Geol. Soc. Amer.* 46 : 245–252.

BIGARELLA, JOAO, and SONIA S. FREIRE. 1960. Nota sóbre a ocorrencia de cascalheiro marinho no Litoral do Paraná. *Boletin da Universidade do Paraná (Brazil) Geologia* No. 3. Inst. de Geologia.

BIRD, J. B. 1959. Recent contributions to the physiography of northern Canada. *Zeit. für Geomorphologie.* 3 (2) : 151–174. Also *in* Problems of the Pleistocene and Arctic. *Pub. McGill Univ. Museums* 1 : 96–117.

BLAKE, S. F. 1953. The Pleistocene fauna of Wailes Bluff and Langley Bluff, Maryland. *Smithsonian Misc. Coll.* 121 (12) : 1–32.

BLAKE, WESTON, JR. 1961. Radiocarbon dating of raised beaches in Nordaustlandet, Spitzbergen. *In* Geology of the Arctic, 1st Internat. Symposium on Arctic Geology 1 : 133–145, Calgary.

BLANC, ALBERT C. 1937. Low levels of the Mediterranean Sea during the Pleistocene glaciation. *Quart. Jour. Geol. Soc. London* 93 : 621–651.

BLEACKLEY, D., and R. A. DUJARDIN. 1959. The Corentyne series of British Guiana. Quatrième Conférence Géologique des Guyanes. *Memoir Carte Géol. France*, 111–116.

BLOOM, ARTHUR L. 1960. Late Pleistocene changes of sea level in southwestern Maine. Maine Geol. Survey, Augusta, Maine.

—— 1961. Evidence of submergence from Connecticut tidal marshes (abstract). *Bull. Geol. Soc. Amer.* Abstracts for 1961 : 138.

BØGGILD, O. B. 1928. The geology of Greenland. *In* Greenland 1 : 231–255, edited by M. Valh *et al.*, Comm. for Direction of Geol. and Geog. Invest. in Greenland. Copenhagen.

BONIFAY, EUGENE, and PAUL MARS. 1959. Le Tyrrhénien dans le cadre de la chronologie quaternaire méditerranéenne. *Bull. Soc. Géol. de France*, Ser. 7, 1 : 62–78.

BOSWORTH, T. O. 1922. Geology of the Tertiary and Quaternary periods in the north-west part of Peru. London, Macmillan.

BOWMAN, I. 1916. The Andes of southern Peru. New York, Amer. Geog. Soc.

BRADLEY, WILLIAM C. 1956. Carbon-14 date for a marine terrace at Santa Cruz, California. *Bull. Geol. Soc. Amer.* 67 : 675–678.

BRETZ, J. HARLAN. 1935. Physiographic studies in east Greenland. *In* The Fiord region of east Greenland, by Louise Boyd. *Amer. Geographic Soc. Special Pub.* 18.

BRODNIEWICZ, IRENA. 1960. Eemskie mieczaki irskie z wiercenia w Brachlewie. (Eemian marine mollusks from a boring in Brachlewo (Poland).) *Acta Paleont. Polonica* 5 (2) : 235–282.

BROECKER, WALLACE. 1961. Radiocarbon dating of late Quaternary deposits, south Louisiana: a discussion. *Bull. Geol. Soc. Amer.* 72 : 159–162.

BROECKER, W. S., and J. L. KULP. 1957. Lamont natural radiocarbon measurements IV. *Science* 126 : 1324–1334.

BROECKER, WALLACE S., MAURICE EWING, and BRUCE HEEZEN. 1960. Evidence for an abrupt change in climate close to 11,000 years ago. *Amer. Jour. Sci.* 258 : 429–448.

BRØGGER, W. C. 1900–1901. Om de senglaciale og postglaciale nivaforandringer i Kristianiafeltet. *Norges Geol. Undersögelse* No. 31.

BROGGI, J. A. 1946. Las terrazas marinas de la Bahia de San Juan. *Bol. Sociedad Geol. del Peru* 19 : 21–33.

BROTZEN, F. 1961. An interstadial (radiocarbon dated) and the substages of the last glaciation in Sweden. *Geol. Fören. Förhandl.*, Stockholm 83 (2) : 144–150.

BROUWER, AART. 1956. Pleistocene transgressions in the Rhine Delta. *Quaternaria* 3 : 83–93.

BRÜGGEN, JUAN. 1950. Fundamentos de la geología de Chile. Santiago, Chile, Instituto Geografico Militar.

CASTANY, H., and F. OTTMANN. 1957. Le Quaternaire marin de la Méditerranée occidentale. *Revue de Geog. physique et Géol. Dynamique*, Sér. 2, 1: 46–55.

CHARLESWORTH, J. K. 1957. The Quaternary era. 2 v. London, Edward Arnold.

CHATWIN, C. P. 1961. British regional geology. East Anglia and adjoining areas. Fourth edition. London, Geol. Surv. and Museum.

CHESTER, FREDERICK D. 1884. The Quaternary gravels of northern Delaware. *Amer. Jour. Sci.*, Ser. 3, 27: 189–199.

—— 1885. Gravels of southern Delaware. *Amer. Jour. Sci.*, Ser. 3, 29: 36–44.

CLAPP, FREDERICK G. 1907. Complexity of the glacial period in northeastern New England. *Bull. Geol. Soc. Amer.* 18: 505–556.

CLARK, W. B. 1906. Mollusca. *In* Pliocene and Pleistocene of Maryland, by George B. Shattuck. Maryland Geol. Surv.

CLARK, W. B., and B. L. MILLER. 1912. Physiography and geology of the coastal plain province of Virginia. *Virginia Geol. Surv. Bull.* 4: 13–122.

COLEMAN, A. P. 1926. The Pleistocene of Newfoundland. *Jour. Geol.* 34: 193–203.

—— 1927. Glacial and interglacial periods in eastern Canada. *Jour. Geol.* 35: 385–403.

—— 1941. The last million years. A history of the Pleistocene in North America. Toronto Univ. Press.

COOKE, C. WYTHE. 1925. The coastal plain. *In* Physical geography of Georgia, *Georgia Geol. Surv. Bull.* 42.

—— 1930a. Correlation of coastal terraces. *Jour. Geol.* 38: 578–589.

—— 1930b. Pleistocene seashores. *Jour. Wash. Acad. Sci.* 20: 389–395.

—— 1931. Seven coastal terraces in the southeastern states. *Jour. Wash. Acad. Sci.* 21: 503–513.

—— 1932. Tentative correlation of American glacial chronology with the marine time scale. *Jour. Wash. Acad. Sci.* 22: 310–312.

—— 1935. Tentative ages of Pleistocene shorelines. *Jour. Wash. Acad. Sci.* 25: 331–333.

—— 1936. Geology of the coastal plain of South Carolina. *U. S. Geol. Surv. Bull.* 867.

—— 1937. The Pleistocene Horry clay and Pamlico formation near Myrtle Beach, South Carolina. *Jour. Wash. Acad. Sci.* 27: 1–5.

—— 1939. Scenery of Florida interpreted by a geologist. *Florida Geol. Surv. Bull.* 17.

—— 1943. Geology of the coastal plain of Georgia. *U. S. Geol. Surv. Bull.* 941.

—— 1945. Geology of Florida. *Fla. Geol. Surv. Bull.* 29.

—— 1952. Sedimentary deposits of Prince Georges County, Maryland, and the District of Columbia. *Maryland Dept. Geology, Mines and Water Resources Bull.* 10: 1–53.

—— 1958. Pleistocene shorelines in Maryland. *Bull Geol. Soc. Amer.* 69: 1187–1190.

COOKE, C. WYTHE, and STUART MOSSON. 1929. Geology of Florida. *Florida Geol. Surv.*, 20th Rept., 29–288.

COULTER, HENRY W., KEITH M. HUSSEY, and JOHN B. O'SULLIVAN. 1960. Radiocarbon dates relating to the Gubik formation, northern Alaska. *U. S. Geol. Surv. Prof. Paper 400-B*: 350–351.

CRAIG, B. G., and J. G. FYLES. 1961. Pleistocene geology of Arctic Canada. *In* Geology of the Arctic, 1st Internat. Symposium on Arctic Geology 1: 403–420. Calgary.

CURRAY, JOSEPH R. 1960. Sediments and history of Holocene transgression, continental shelf, northwest Gulf of Mexico. *In* Recent sediments, northwest Gulf of Mexico. Tulsa, Amer. Assoc. Petrol. Geol.

—— 1961. Late Quaternary sea level: A discussion. *Bull. Geol. Soc. Amer.* 72: 1707–1712.

CUSHMAN, JOSEPH A. 1906. The Pleistocene deposits of Sankoty Head, Nantucket and their fossils. *Nantucket Maria Mitchell Assoc.* 1 (1): 1–21.

DACHNOWSKI-STOKES, A. P., and B. W. WELLS. 1929. The vegetation, stratigraphy and age of the "Open Land" peat area in Cartaret county, North Carolina. *Jour. Wash. Acad. Sci.* 19: 1–11.

DALL, WILLIAM H. 1890–1903. Contributions to the Tertiary fauna of Florida. *Trans. Wagner Free Inst. Sci.* 3 (I–VI).

—— 1903. Synopsis of the family Astartidae, with a review of the American species. *Proc. U. S. Natl. Mus.* 26 (1342): 933–951.

—— 1912. New species of fossil shells from Panama and Costa Rica. *Smithson. Misc. Coll.* 59 (2): 1–10.

—— 1920. Pliocene and Pleistocene fossils from the Arctic coast of Canada and the auriferous beaches of Nome, Norton Sound, Alaska. *U. S. Geol. Surv. Prof. Paper 125.*

DALY, R. A. 1902. The geology of the northeast coast of Labrador. *Bull. Mus. Comp. Zool.* 38: 205–270.

—— 1934. The changing world of the ice age. New Haven, Yale Univ. Press.

DARWIN, CHARLES. 1846. Geological observations on South America. Being the third part of the Geology of the voyage of the "Beagle" . . . during 1832 to 1836. London.

DAVIS, W. M. 1933. Glacial epochs of the Santa Monica Mountains, California. *Bull. Geol. Soc. Amer.* 44: 1041–1133.

DAWSON, J. W. 1871. Notes on the post-Pliocene geology of Canada. *Canadian Natur. Quart. Jour. Sci., n.s.* 6: 19–42, 166–187, 251–259, 369–416. Reprint (1872), 1–122.

DEEVEY, EDWARD S. 1950. Hydroids from Louisiana and Texas, with remarks on the Pleistocene biogeography of the western Gulf of Mexico. *Ecology* 31 (3): 334–367.

DÉPÉRET, C. 1906. Les anciennes lignes de rivage de la côte française de la Méditerranée. *Bull. Soc. Géol. France*, Paris (4) 6: 207–230.

DESOR, EDOUARD. 1849. On the Tertiary and more recent deposits in the island of Nantucket. *Quart. Jour. Geol. Soc. London* 5: 340–344.

DIXON, C. G. 1955. Notes on the geology of British Honduras. Belize, B. H.

DOERING, JOHN A. 1956. Review of Quaternary surface formations of Gulf Coast region. *Bull. Amer. Assoc. Petrol. Geol.* 40: 1816–1862.

—— 1958. Citronelle age problem. *Bull. Amer. Assoc. Petrol. Geol.* 42: 764–786.

—— 1960. Quaternary surface formations of southern part of Atlantic coastal plain. *Jour. Geol.* 68: 182–202.

DONN, WILLIAM, WILLIAM FARRAND, and MAURICE EWING. 1962. Pleistocene ice volumes and sea level lowering. *Jour. Geol.* 70: 206–214.

DONNER, J. J. 1959. The late and post glacial raised beaches in Scotland. *Ann. Acad. Sci. Fennicae*, Ser. A. III, 53. Helsinki.

DONNER, J. J., and R. G. WEST. 1957. The Quaternary geology of Brageneset, Nordaustlandet, Spitzbergen. *Norsk Polar-institutt, Skrifter* 109.

DREIMANIS, ALEKSIS. 1947. A draft of Pleistocene stratigraphy in Latvia and south Estonia. *Geol. Förn Förhandl.* 69 (4): 465–470. Stockholm.

—— 1949. Interglacial deposits of Latvia. *Geol. Förn. Förhandl.* 71, 4: 525–536.

—— 1960. Pre-classical Wisconsin in the eastern portion of the Great Lakes region, North America. 21st Internat. Geol. Cong., Pt. 4: 108–119. Copenhagen.

DUBAR, JULES. 1958a. Stratigraphy and paleontology of the late Neogene strata of the Caloosahatchee River area of southern Florida. *Florida Geol. Surv. Bull.* 40.

—— 1958b. Neogene stratigraphy of southwestern Florida. *Trans. Gulf Coast Assoc. Geol. Soc.* **8**: 129–155.

DUBOIS, G. 1924. Recherches sur les terraines quaternaires du nord de la France. *Mémoir Soc. Géol. Nord., Lille,* **8**: 1–356.

DUNBAR, M. J. 1959. Arctic marine zoogeography. *In* Problems of the Pleistocene and Arctic, *Pub. McGill Univ. Museums* **1**: 55–63.

DURHAM, J. WYATT, and EDWIN C. ALLISON. 1960. The geologic history of Baja California and its marine faunas. *Systematic Zool.* **9**: 47–91.

EMERSON, WILLIAM K. 1960. Pleistocene invertebrates from near Punta San Jose, Baja California, Mexico. *Amer. Mus. Nat. Hist. Novitates* **2002**.

EMERSON, WILLIAM K., and WARREN O. ADDICOTT. 1958. Pleistocene invertebrates from Punta Baja, Baja California, Mexico. *Amer. Mus. Nat. Hist. Novitates* **1909**: 1–11.

EMERY, K. O. 1958. Shallow submerged marine terraces of southern California. *Bull. Geol. Soc. Amer.* **69**: 39–60.

—— 1960. The sea off southern California. New York, Wiley and Sons.

EMILIANI, CAESARE. 1955. Pleistocene temperatures. *Jour. Geol.* **63**: 538–578.

EMILIANI, C., T. MAYEDA, and R. SELLI. 1961. Paleotemperature analysis of the Plio-Pleistocene section at La Castella, Calabria, Italy. *Bull. Geol. Soc. Amer.* **72**: 679–688.

ERICSON, DAVID, MAURICE EWING, GOESTA WOLLIN, and BRUCE HEEZEN. 1961. Atlantic deep sea sediment cores. *Bull. Geol. Soc. Amer.* **72**: 193–286.

EWING, M., J. EWING, and C. FRAY. 1960. Buried erosional terrace on the edge of the continental shelf east of New Jersey. *Bull. Geol. Soc. Amer.* **71**: 1860.

EWING, MAURICE, CHARLES FRAY, and E. DAHLBERG. 1960. Sediments from Argentina continental shelf: preliminary report (abstract). *Bull. Geol. Soc. Amer.* **71**: 2094.

FAIRBRIDGE, RHODES W. 1958. Dating the latest movements of Quaternary sea level. *Trans. N. Y. Acad. Sci.* **20**: 471–482.

—— 1961. Sea level and the Holocene boundary in the eastern United States. 6th INQUA congress, Warsaw, Poland, Sept., 1961. Abstract, 187.

FARRAND, WILLIAM R. 1961. Frozen mammoths and modern geology. *Science* **133** (3455): 729–735.

FARRAND, WILLIAM R., and R. T. GAJDA. 1961. Isobases of maximum post-glacial submergence in glaciated North America, east of the Cordillera. 6th INQUA Congress in Warsaw, Poland, in September, 1961. Abstract, 178.

—— 1962. Isobases of the Wisconsin marine limit in Canada. *Geogr. Bull.* **17**, Canadian Dept. Mines and Techn. Surveys.

FEDEROV, P. V. 1960. Stratigrafiya Chetvertichn'x otlojenii Ponto-Kaspiya. (Stratigraphy of Quaternary deposits of Ponto-Caspian.) Internat. Geol. Cong. 21st sess. Rept. of Soviet Geologists, Problem 4 "Chronology and Climatology of the Quaternary," 48–57. In Russian with English summary. Moscow.

—— 1961. Biostratigrafia Chetvertichyx morskix otlojenii Ponto-Kaspiiskoi oblasti. (Biostratigraphy of Quaternary marine deposits of the Ponto-Caspian Region.) *Dokladi Sovetskix Geologov;* VI Kongress INQUA, 96–106. Moscow.

FEDEROV, P. V., and L. A. SKIBA. 1960. Kolebania urovnei Chernogo i Kaspiiskogo Morei b Golocene. (Variations of levels of the Black and Caspian Seas during the Holocene.) *Isvestia Akad. Nauk U.S.S.R. Seria Geograficheskaya* **4**: 24–34.

FERUGLIO, EGIDIO. 1933. I Terrazzi marini della Patagonia. *Giornale di Geologia,* Bologna **8**: 1–28.

—— 1948. Edad de las terrazas marinas de la Patagonia. 18th Internat. Geol. Cong., Pt. 9: 30–39. London.

—— 1950. Descripción geológica de la Patagonia. (3 v.) Dirección General de Yacimentos Petrolíferos Fiscales, Buenos Aires.

FEYLING-HANSSEN, ROLF W. 1955. Stratigraphy of the marine late-Pleistocene of Billefjorden, Vestspitsbergen. *Norsk Polarinstitutt Skrifter* **107**. Oslo.

FEYLING-HANSSEN, R. W., and F. A. JORSTAD. 1950. Quaternary fossils from the Sassen-area in Isfjorden, West-Spitzbergen. *Norsk Polarinstitutt Skrifter* **94**. Oslo.

FISK, H. N. 1938. Geology of Grant and LaSalle parishes. *Louisiana Dept. Consv. Geol. Bull.* **10**: 5–246.

—— 1944. Geological investigations of the alluvial valley of the lower Mississippi River, 1–78. Mississippi River Comm. Vicksburg.

FISK, H. N., and E. McFARLAN. 1955. Late Quaternary deltaic deposits of the Mississippi River. *In* The crust of the earth, *Geol. Soc. Amer. Special Paper* **62**: 279–302.

FLINT, RICHARD F. 1940. Late Quaternary changes of level in western and southern Newfoundland. *Bull. Geol. Soc. Amer.* **51**: 1757–1780.

—— 1940b. Pleistocene features of the Atlantic coastal plain. *Amer. Jour. Sci.* **238**: 757–787.

—— 1942. Atlantic coastal plain "terraces." *Jour. Wash. Acad. Sci.* **32**: 235–237.

—— 1948a. Studies on glacial geology and geomorphology. *In* The coast of northeast Greenland, by Louise Boyd. *Amer. Geographic Soc. Special Pub.* **30**.

—— 1948b. Glacial and Pleistocene geology. New York, John Wiley.

FLINT, RICHARD F., *et al.* 1945. Glacial map of North America. *Geol. Soc. Amer. Special Paper* **60**.

FLINT, RICHARD F., and EDWARD S. DEEVEY. 1951. Radiocarbon dating of late Pleistocene events. *Amer. Jour. Sci.* **249**: 257–300.

FLINT, RICHARD FOSTER, and FRIEDERICH BRANDTNER. 1961. Climatic changes since the Interglacial. *Amer. Jour. Sci.* **259**: 321–328.

FROMM, ERIK. 1953. Nedisning och landhöjning under Kvärtartiden. (Glaciation and changes of level in Quaternary Age.) *Atlas över Sverige,* No. 19–20. Stockholm, Svenska Sällskapet för Antropoligi och Geografi.

FRYE, JOHN C., and H. B. WILLMAN. 1960. Classification of the Wisconsin stage in the Lake Michigan glacial lobe. *Illinois Geol. Surv. Circular* **285**.

FULLER, M. L. 1914. The geology of Long Island, New York. *U. S. Geol. Surv. Prof. Paper* **82**.

GADD, N. R. 1961. Surficial geology of the Ottawa area. Report of progress. *Geol. Surv. Canada Paper* **61-19**.

GALON, RAJMUND, *et al.* 1961. Guide-book of excursion from the Baltic to the Tatras. Part 1, North Poland. 6th Congress of INQUA (Internat. Assoc. for Quaternary Research), Poland.

GARDNER, JULIA. 1943. Mollusca from the Miocene and lower Pliocene of Virginia and North Carolina. Part 1. Pelecypoda. *U. S. Geol. Surv. Prof. Paper* **199-A**.

—— 1948. Mollusca from the Miocene and lower Pliocene of Virginia and North Carolina. Part 2. Scaphopoda and gastropoda. *U. S. Geol. Surv. Prof. Paper* **199-B**.

GERASIMOV, I. P., and K. K. MARKOV. 1939. The glacial period in the territory of U.S.S.R. *U.S.S.R. Acad. Sci. Inst. Geog. Trans.* **33**:443–462. In Russian with English summary.

GIGNOUX, M. 1913. Les formations marines pliocènes et quaternaires de l'Italie du sud de la Sicile. *Ann. Univ. Lyon,* n.s. **1**(36).

GIGOUT, MARCEL. 1956. Réponse au questionaire de la commission des lignes de rivage du IV congrès international de l'INQUA (1953). *Quaternaria* **3**: 71–79.

GOLDRING, WINIFRED. 1922. The Champlain Sea. *Rept. Director N. Y. State Mus. for 1920–21*: 153–194.

GOLDTHWAIT, J. W. 1924. Physiography of Nova Scotia. *Geol. Surv. Canada. Memoir* **140**.

GOLDTHWAIT, R. P. 1958. Wisconsin age forests in western Ohio. I. Age and glacial events. *Ohio Jour. Sci.* **58**: 209–230.

GRANT, U. S., IV, and H. R. GALE. 1931. Catalogue of the marine Pliocene and Pleistocene mollusca of California and adjacent regions. *San Diego Soc. Nat. Hist. Memoir* 1.

GREELY, A. W. 1888. Report on the proceedings of the United States expedition to Lady Franklin Bay, Grinnell Bay (Internat. Polar Exped.). *House Misc. Documents* 2427, 2428.

GROMOV, V. 1945. Twenty-five years of study of the Quaternary in the U.S.S.R. *Amer. Jour. Sci.* 243: 492–516.

GUNTER, HERMAN, and G. M. HANNA. 1933. Notes on the geology and occurrence of some diatomaceous earth deposits of Florida. *Florida Geol. Surv.*, 23–24th *Ann. Rept.*, 57–64.

HACK, JOHN T. 1955. Geology of the Brandywine area and origin of the Upland of southern Maryland. *U. S. Geol. Surv. Prof. Paper* 267-A.

HAFSTEN, ULF. 1958. De senkvartaere strandlinjeforskyvninge i Oslotrakten belyst ved pollenanalytiske undersøkelser. (Application of pollen analysis in tracing late Quaternary displacement of shorelines in the inner Oslofjord area.) *Norsk Geografisk Tidsskrift* B XVI, Hefte 1–8, 1957–1958: 74–99.

HALICKI, B. 1950. Zagadnien stratygrafii Plejstocenu na Niza Europejskim. (Some problems concerning the stratigraphy of the European Lowland.) *Acta Geol. Polonica* 1(2): 106–142.

HAMMEN, THOMAS VAN DER. 1959. First results of pollen analysis in British Guiana. 5th Inter-Guiana Geol. Conference, Georgetown, 1–5.

HANNA, G. DALLAS. 1933. Diatoms of the Florida peat deposits. *Florida Geol. Surv.*, 23–24th *Ann. Rept.*, 69–119.

HANSEN, SIGURD, and ARNE VAGN NIELSEN. 1960. Glacial geology of southern Denmark. 21st Internat. Geol. Congress Guidebook, Copenhagen.

HANSON, GEORGE. 1934. The Bear River Delta, British Columbia, and its significance regarding Pleistocene and Recent glaciation. *Trans. Royal Soc. Canada*, Ser. 3, 28, Sec. 4: 179–185.

HERTLEIN, LEO G., and WILLIAM K. EMERSON. 1956. Marine Pleistocene invertebrates from near Puerto Penasco, Sonora, Mexico. *Trans. San Diego Soc. Nat. Hist.* 12(8): 154–176.

HESSLAND, IVAR. 1945. On the Quaternary *Mya* period in Europe. *Arkiv för Zoologi* 37(8): 1–51. Stockholm.

HEUSSER, CALVIN J. 1960. Late Pleistocene environments of North Pacific North America. *Amer. Geogr. Soc. Special Pub.* 35.

HOLMES, FRANCIS S. 1860. Post-Pleiocene fossils of South-Carolina. Charleston, S. C., Russell & Jones.

HOPKINS, D. M., F. S. MacNEIL, and E. B. LEOPOLD. 1960. The coastal plain at Nome, Alaska: a late Cenozoic type section for the Bering Strait region. 21st Internat. Geol. Cong. Pt. 4: 46–57. Copenhagen.

HORBERG, C. L. 1955. Radiocarbon dates and Pleistocene chronological problems in the Mississippi Valley region. *Jour. Geol.* 63: 278–285.

HÖRNER, NILS. 1929. Late glacial marine limit in Massachusetts. *Amer. Jour. Sci.* 17: 123–145.

HOWELL, B. F., and HORACE G. RICHARDS. 1937. The fauna of the "Champlain Sea" of Vermont. *Nautilus* 51: 8–10.

HUBBARD, BELA. 1923. The geology of the Lares district. *New York Acad. Sci. Scient. Surv. Porto Rico* 3(2): 79–164.

HUBBS, CARL L., GEORGE S. BIEN, and HANS E. SUESS. 1960. La Jolla natural radiocarbon measurements. *Amer. Jour. Sci. Radiocarbon Suppl.* 2: 197–223.

HYYPPÄ, ESA. 1955. On the Pleistocene geology of southeastern New England. *Bull. Comm. Geol. de Finlande* 167: 153–225.

ILLIES, HENNING. 1960. Geologie der gegend von Valdivia, Chile. *Neues Jahrbuch für Geologie und Paläontologie*, Abh. Bd. 111, Heft 1: 30–110. Stuttgart.

IVES, R. L. 1951. High sea levels of the Sonoran shore. *Amer. Jour. Sci.* 249: 215–223.

JELGERSMA, S., and A. J. PANNEKOEK. 1960. Post-glacial rise of sea level in the Netherlands (a preliminary review). *Geologie en Mijnbouw* 39(6): 201–207.

JENNESS, STUART E. 1960. Late Pleistocene glaciation of eastern Newfoundland. *Bull. Geol. Soc. Amer.* 71: 161–180.

JOHNSON, B. L. 1907. Pleistocene terracing in North Carolina coastal plain. *Science, n.s.,* 26: 640–642.

JOHNSON, C. W. 1934. List of marine mollusca of the Atlantic coast from Labrador to Texas. *Proc. Boston Soc. Nat. Hist.* 40(1): 1–204.

JOHNSON, RALPH G. 1962. Mode of formation of marine fossil assemblages of the Pleistocene Millerton formation of California. *Bull. Geol. Soc. Amer.* 73: 113–130.

JOHNSTON, W. A. 1916. Late Pleistocene oscillations of sea-level in the Ottawa Valley. *Canadian Geol. Surv. Museum Bull.* 24 (Geological Series 33): 1–14.

JUDSON, SHELDON. 1949. The Pleistocene stratigraphy of Boston, Massachusetts, and its relation the Boyleston Street fishwier. *In* The Boyleston Street fishwier, by Frederick Johnson *et al., Papers of the Peabody Foundation for Archaeology* 4(1): 7–48.

KANAKOFF, GEORGE P., and WILLIAM K. EMERSON. 1959. Late Pleistocene invertebrates of the Newport Bay area, California. *Los Angeles County Museum Contrib. in Science* 31: 1–47.

KAYE, CLIFFORD A. 1959. Shoreline features and Quaternary shoreline changes, Puerto Rico. *U. S. Geol. Surv. Prof. Paper* 317-B.

—— 1961. Pleistocene stratigraphy of Boston, Massachusetts. *U. S. Geol. Surv. Prof. Paper* 424-B: 73–76.

KINDLE, E. M. 1924. Geology of a portion of the northern part of the Moose River Basin, Ontario. *Canadian Geol. Surv. Summ. Rept. for 1923* 101: 21–41.

KNIGHT, J. BROOKS. 1933. *Littorina irrorata*, a post-Pleistocene fossil in Connecticut. *Amer. Jour. Sci.*, Ser. 5, 26: 130–133.

KOETTLITZ, REGINALD. 1898. Observations on the geology of Franz Josef Land. *Quart. Jour. Geol. Soc. London* 54: 620–645.

LAMOTHE, R. DE. 1911. Les anciennes lignes de rivage du Sahel d'Alger et d'une partie de la côte algérienne. *Mémoir Soc. géol. France* (4) 1(6): 1–288.

LAROCQUE, A. 1949. Post-Pleistocene connection between James Bay and the Gulf of St. Lawrence. *Bull. Geol. Soc. Amer.* 60: 363–380.

—— 1953. Catalogue of the Recent Mollusca of Canada. *Nat. Mus. of Canada Bull.* 129.

LAURSEN, DAN. 1950. The stratigraphy of the marine deposits in west Greenland. *Meddel. øm Gronland* 151(1).

LAVROVA, M. A. 1960. Chetvertichnaya geologiya Kol'skogo polyostrova. (Quaternary geology of the Kola peninsula.) Moscow-Leningrad, Acad. Nauk U.S.S.R.

LAVROVA, M. A., and S. L. TROITSKY. 1960. Mejlednikob'e transgressii na Severe Evropi i Sibiri. (Interglacial transgressions in northern Europe and Siberia.) Internat. Geol. Congress, 21st sess., Sect. 4, Chronology and climatology of the Quaternary, 124–136. In Russian with English summary. Moscow.

LEAVITT, H. WALTER, and EDWARD H. PERKINS. 1935. Glacial geology of Maine. *Maine Technology Experiment Station Bull.* 30.

LECOINTRE, GEORGES. 1952. Recherches sur le Néogène et le Quaternaire marins de la Côte Atlantique du Maroc. *Div. des Mines et de la Géologie, Maroc. Notes et Mémoires,* No. 99 (2 v.).

LEE, HURLBERT A. 1960. Late glacial and postglacial Hudson Bay episode. *Science* 131 (3412): 1609–1611.

LEIGHTON, MORRIS M. 1960. The classification of the Wisconsin glacial stage of north central United States. *Jour. Geol.* 68: 529–552.

LEMON, R. R. H., and C. S. CHURCHER. 1961. Pleistocene geology and paleontology of the Talara region, northwest Peru. *Amer. Jour. Sci.* **259**: 410–429.

LIDDLE, R. A. 1946. Geology of Venezuela and Trinidad. Ithaca, N. Y., Paleont. Res. Inst.

LINDAL, J. H. 1939. The interglacial formation in Vioidalur, northern Iceland. *Quart. Jour. Geol. Soc. London* **95**: 261–273.

LOUGEE, RICHARD J. 1958. Champlain marine stage at Cochrane, Ontario (abstract). *Bull. Geol. Soc. Amer.* **69**: 1764.

MACCLINTOCK, PAUL. 1940. Weathering of the Jerseyan till. *Bull. Geol. Soc. Amer.* **51**: 103–116.

MACCLINTOCK, PAUL, and HORACE G. RICHARDS. 1936. Correlation of late Pleistocene marine and glacial geology of New Jersey and New York. *Bull. Geol. Soc. Amer.* **47**: 289–338.

MACCLINTOCK, PAUL, and W. H. TWENHOFEL. 1940. Wisconsin glaciation of Newfoundland. *Bull. Geol. Soc. Amer.* **51**: 1729–1756.

MACCLINTOCK, PAUL, and J. TERASMAE. 1960. Glacial history of Covey Hill. *Jour. Geol.* **68**: 232–241.

MCFARLAN, E. 1961. Radiocarbon datings of late Quaternary deposits, south Louisiana. *Bull. Geol. Soc. Amer.* **72**: 129–158.

MACNEIL, F. STEARNS. 1938. Species and genera of Tertiary Noetinae. *U. S. Geol. Surv. Prof. Paper* **189-A**.

MACNEIL, F. S., J. B. MERTIE, and H. A. PILSBRY. 1943. Marine invertebrate faunas of the buried beaches near Nome. *Jour. Paleont.* **17**: 69–96.

MADSEN, VICTOR, V. NORDMAN, and N. HARTZ. 1908. Eemzonerne. Studier over cyprinareret og andere Eem-afferinger i Denmark, Nord Tyskland og Holland (résumé en français). *Danmarks geologiske Undersøgelse*, ser. 2, no. 17.

MANSFIELD, W. C. 1928. Notes on Pleistocene faunas of Maryland, and Pliocene and Pleistocene faunas from North Carolina. *U. S. Geol. Surv. Prof. Paper* **150-F**.

MASON, RONALD J. 1960. Early man and the age of the Champlain Sea. *Jour. Geol.* **68**: 366–376.

MATLEY, C. A. 1926. The geology of the Cayman Islands (British West Indies) and their relation to the Bartlett Trough. *Jour. Geol. Soc. London* **82**: 352–387.

MATSON, G. S., and S. SANFORD. 1913. Geology and ground water of Florida. *U. S. Geol. Surv. Water Supply Paper* 319.

MERCER, J. H. 1956. Geomorphology and glacial history of southernmost Baffin Island. *Bull. Geol. Soc. Amer.* **67**: 553–570.

MITCHELL, G. F. 1960. The Pleistocene history of the Irish Sea. *Adv. of Science* **17**: 313–325.

MITCHELL, G. J. 1922. Geology of the Ponce area. *New York Acad. Sci. Scient. Surv. Porto Rico* **1**(3): 229–300.

MOORE, WAYNE. 1956. Stratigraphy of Pleistocene terrace deposits in Virginia (abstract). *Bull. Geol. Soc. Amer.* **67**: 755.

MOURANT, A. E. 1933. The raised beaches and other terraces of the Channel Islands. *Geol. Mag.* **70**: 58–66.

—— 1935. The Pleistocene deposits of Jersey. *Bull. Soc. Jersey* **12**: 489–496.

MURRAY, GROVER. 1961. Geology of the Atlantic and Gulf coastal province of North America. New York, Harpers.

NANSEN, F. 1910. Greenland. *In* Encyl. Britannica, 11th edit. **12**: 542–548.

NEVESSKAYA, L. A. 1958. Smena kompleksov dvustvorchat'x mollioskov Chernogo Morya v Pozdnechetvertichnoe vremya. (The sequence of pelecypod assemblages of the Black Sea in late Quaternary time.) *Dokladi Akad. Nauk U.S.S.R.* **121**(1): 152–154.

NEVESSKAYA, L. A., and E. N. NEVESSKII. 1961. O sootnoshenii Karangatskix i Novoevksinskix sloev v pribrejn'x raionax Chernogo Morya. (On the correlation of the Karangatian and New Euxinian Beds in the littoral region of the Black Sea.) *Dokladi Akad. Nauk U.S.S.R.* **137**:(4): 934–937.

NEWELL, NORMAN. 1960. Marine planation of tropical limestone islands. *Science* **132**: 144–145.

NEWMAN, WALTER S., and RHODES W. FAIRBRIDGE. 1960. Glacial lakes in Long Island Sound. *Bull. Geol. Soc. Amer.* **71**: 1936.

NICHOLS, D. A. 1936a. Physiographic studies in the eastern Arctic. *Canadian Surveyor* **5**: 2–7.

—— 1936b. Post-Pleistocene fossils of the uplifted beaches of the eastern Arctic regions of Canada. *Canadian Field-Natur.* **50**: 127–129.

NICHOLS, ROBERT L., and STINSON, G. LORD. 1937. Fossiliferous eskers and outwash plains. *Proc. Geol. Soc. Amer.* for 1937: 324–325.

NORDMAN, V. 1928. La position stratigraphique des depôts d'Eem. *Danmarks geologiske Undersøgelse*, Ser. 2, No. 47.

NORDMAN, V., *et al.* 1928. Summary of the geology of Denmark. *Danmarks geologiske Undersøgelse*, Ser. 5, No. 4.

NOTA, D. J. G. 1958. Sediments of the western Guiana shelf. *Mededelingen van de Landbouw-Hogeschool te Wageningen* (Nederland) **58**(2).

OCKELMANN, W. K. 1958. The zoology of east Greenland. Marine lamellibranchiata. *Meddel. om Grønland* **122**(4).

OLSON, E. A., and W. S. BROECKER. 1959. Lamont natural radiocarbon measurements V. *Amer. Jour. Sci.* **257**: 1–16.

OLSSON, A. A. 1940a. Tertiary deposits of northwestern South America and Panama. *Proc. 8th American Sci. Cong.* **4**: 231–287. Washington.

—— 1940b. Some tectonic interpretations of the geology of northwestern South America. *Proc. 8th American Sci. Conference* **4**: 401–416. Washington.

OLSSON, AXEL A., and ANNE HARBISON. 1953. Pliocene mollusca of southern Florida. *Acad. Nat. Sci. Phila. Monograph* 8.

O'NEILL, J. J. 1924. The geology of the Arctic coast of Canada west of the Kent peninsula. *Rept. Canadian Arctic Exped. 1913–1918* **11**(A).

ORR, PHIL C. 1960. Late Pleistocene marine terraces on Santa Rosa Island, California. *Bull. Geol. Soc. Amer.* **71**: 1113–1119.

OURVANTZEV, N. 1930. Die Quätare vereisung des Taimyr-Gebiets. *Zeit. fur Gletscherkunde* **18**: 337–345.

PACKARD, A. S. 1865. Observations on the glacial phenomena of Labrador and Maine, and a view of the Recent invertebrate fauna of Labrador. *Memoir Boston Soc. Nat. Hist.* **1**: 210–303.

PARKER, G. G., and C. WYTHE COOKE. 1944. Late Cenozoic geology of southern Florida with a discussion of the groundwater. *Florida Geol. Surv. Bull.* **27**: 1–119.

PARKER, ROBERT H. 1960. Ecology and distributional patterns of marine macro-invertebrates, northern Gulf of Mexico. *In* Recent sediments northwest Gulf of Mexico, 302–337. Tulsa, Amer. Assoc. Petrol. Geol.

PORTA, J. DE, and N. SOLE DE PORTA. 1960. El Cuaternario marino de la isla de Tierrabomba (Bolívar). *Boletin de Geología* (Univ. Santander, Bucarmanga, Colombia) **4**: 19–44.

POTTER, DAVID. 1932. Botanical evidence for a post-Pleistocene connection between Hudson Bay and the St. Lawrence Basin. *Rhodora* **34**: 68–89; 101–112.

POWERS, SIDNEY. 1918. Notes on the geology of eastern Guatemala and northwestern Spanish Honduras. *Jour. Geol.* **26**: 507–523.

PRICE, W. ARMSTRONG. 1933. Role of diastrophism in topography of Corpus Christi area, south Texas. *Bull. Amer. Assoc. Petrol. Geol.* **17**: 907–962.

—— 1934. Lissie formation and Beaumont clay in south Texas. *Bull. Amer. Assoc. Petrol. Geol.* **18**: 948–959.

—— 1956. Environment and history in identification of shoreline types. *Quaternaria* **3**: 151–166. Rome.

—— 1958. Sedimentology and Quaternary geomorphology of south Texas. *Trans. Gulf Coast Assn. Geol. Soc.* **8**: 41–75.

PRICE, W. ARMSTRONG, and THEODORE D. COOK, et al. 1958. Field trip guidebook. Gulf Coast Ass'n Geol. Soc., Cambe Log Co., Houston, Texas.

PUGH, G. T. 1906. Pleistocene deposits of South Carolina. Thesis, Vanderbilt Univ., Nashville, Tenn.

PURI, HARBANS, and ROBERT O. VERNON. 1959. Summary of the geology of Florida and a guidebook to the classic exposures. Florida Geol. Surv. Special Pub. 5.

RAMSAY, WILHELM. 1904. Beiträge zur geologie der recenten und Pleistocänen bildungen der Halbinsel Kanin. Fennia 21(7) : 1–66. Helsinki.

—— 1930. Changes of sea level resulting from the increase and decrease of glaciations. Fennia 52(5) : 1–62. Helsinki.

RASMUSSEN, W. C., and T. H. SLAUGHTER. 1955. The ground water resources (of Somerset, Wicomico and Worcester counties, Maryland). Maryland Dept. Geology, Mines and Water Resources. Bull. 16: 1–170.

RASMUSSEN, W. C., and T. H. SLAUGHTER. 1957. The ground water resources (of Caroline, Dorchester and Talbot counties, Maryland). Maryland Dept. Geology, Mines and Water Resources Bull. 18: 1–372.

RENZ, H. H. 1940. Stratigraphy of northern South America, Trinidad and Barbados. Proc. 8th American Scientific Cong. 4: 513–571.

RICHARDS, HORACE G. 1930. Fossil mollusks and other invertebrates from the Hudson River tube, New York and New Jersey. Nautilus 43: 131–132.

—— 1933. Marine fossils from New Jersey indicating a mild interglacial stage. Proc. Amer. Philos Soc. 72: 181–214.

—— 1935. Pleistocene mollusks from western Cuba. Jour. Paleont. 9: 253–258.

—— 1936a. Fauna of the Pleistocene Pamlico formation of the southern Atlantic coastal plain. Bull. Geol. Soc. Amer. 47: 1611–1656.

—— 1936b. Recent and Pleistocene marine shells of James Bay. Amer. Midl. Natur. 17: 528–545.

—— 1937. Land and freshwater mollusks from the island of Cozumel, Mexico, and their bearing on the geological history of the region. Proc. Amer. Philos. Soc. 77: 249–262.

—— 1938a. Land mollusks from the island of Roatan. Proc. Amer. Philos. Soc. 79: 167–178.

—— 1938b. Marine Pleistocene of Florida. Bull. Geol. Soc. Amer. 49: 1267–1296.

—— 1939a. Marine Pleistocene of the Gulf coastal plain: Alabama, Mississippi and Louisiana. Bull. Geol. Soc. Amer. 50: 297–316.

—— 1939b. Marine Pleistocene of Texas. Bull. Geol. Soc. Amer. 50: 1885–1898.

—— 1940a. Pleistocene fossils from the Belcher Islands in Hudson Bay. Ann. Carnegie Mus. 28: 47–52.

—— 1940b. Marine Pleistocene fossils from Newfoundland. Bull. Geol. Soc. Amer. 51: 1781–1788.

—— 1941. Post-Wisconsin fossils from the west coast of Hudson Bay. Acad. Nat. Sci. Phila. Not. Nat. 84.

—— 1943a. Pliocene and Pleistocene mollusks from the Santee-Cooper area, South Carolina. Acad. Nat. Sci. Phila. Not. Nat. 118.

—— 1943b. Pleistocene mollusks from Margarita Island, Venezuela. Jour. Paleont. 17: 120–123.

—— 1944. Notes on the geology and paleontology of the Cape May canal, New Jersey. Acad Nat. Sci. Phila. Not. Nat. 134.

—— 1945. Subsurface stratigraphy of Atlantic coastal plain between New Jersey and Georgia. Bull. Amer. Assoc. Petrol. Geol. 29: 885–955.

—— 1950. Postglacial marine submergence of Arctic North America with special reference to the Mackenzie Delta. Proc. Amer. Philos. Soc. 94: 31–37.

—— 1954. The Pleistocene of Georgia. Georgia Mineral News Letter 7: 110–114.

—— 1955. The geological history of the Cayman Islands. Acad. Nat. Sci. Phila. Not. Nat. 284.

—— 1959. Recent studies of the Pleistocene of the South Atlantic coastal plain. Southeastern Geol. 1: 11–21.

—— 1960. The dating of the "Subway Tree" of Philadelphia. Proc. Penna. Acad. Sci. 34: 107–108.

RICHARDS, HORACE G., and JAMES L. RUHLE. 1955. Mollusks from a sediment core from the Hudson submarine canyon. Proc. Penna. Acad. Sci. 29: 186–190.

ROYO Y GOMEZ, J. 1959. El glaciarismo Pleistocene en Venezuela. Boletin Informativo, Assoc. Venezolana de Geología, Minera y Petróleo 2, (11) : 333–353. Caracas.

RUGGIERI, G., and R. SELLI. 1948. Il Pliocene e il postpliocene dell'Emilia. 18th Internat. Geol. Cong. Pt. 9: 85–93. London.

RÜHLE, EDWARD, et al. 1961. Mapy z atlasu geologicznego Polski. Instytut Geologiczny, Warsaw.

SACHS, V. N., and S. A. STRELKOV. 1959. Chetvertichnie otlojeniya Sovetskoi Arktiki. (Quaternary geology of the Soviet Arctic.) Trans. Arctic Geol, Research Inst. Ministry of Geol. and Mineral Conserv. 91: 1–232. Moscow.

SACHS, V. N., and S. A. STRELKOV. 1961. Mesozoic and Cenozoic of the Soviet Arctic. In Geology of the Arctic, 1st Internat. Symposium of Arctic Geol. 1: 48–67. Calgary.

SALISBURY, R. D., and G. N. KNAPP. 1917. The Quaternary formations of southern New Jersey. New Jersey Geol. Surv. 8.

SAURAMO, MATTI. 1929. The Quaternary geology of Finland. Bull. Commission Geol. de Finlande 86: 1–110.

—— 1958. Die Geshichte der Ostsee. Ann. Acad. Sci. Fennicae, Ser. A III, No. 51: 1–522. Helsinki.

SCHRADER, F. C. 1904. A reconnaissance in northern Alaska . . . in 1901. U. S. Geol. Surv. Prof. Paper 20.

SCHUCHERT, CHARLES. 1935. Historical geology of the Antillean-Caribbean region. New York, Wiley.

SELLARDS, E. H. 1919. Geological sections across the Everglades of Florida. Florida Geol. Surv. 12th Ann Rept., 71–73.

SHATTUCK, G. B., et al. 1906. The Pliocene and Pleistocene of Maryland. Maryland Geol. Surv.

SHEPARD, FRANCIS P. 1960. Rise of sea level along northwest Gulf of Mexico. In Recent sediments northwest Gulf of Mexico, 338–344. Tulsa, Amer. Assoc. Petrol. Geol.

SHEPPARD, GEORGE. 1937. The geology of south-west Ecuador. London, Thomas Murby and Co.

SHIMER, H. W. 1908. Dwarf faunas. Amer. Natur. 42: 472–490.

—— 1918. Post-glacial history of Boston. Proc. Amer. Acad. Arts and Sci. 53: 439–463.

SINNOTT, ALLEN, and G. CHASE TIBBITTS. 1961. Pleistocene terraces on the Eastern Shore Peninsula, Virginia. U. S. Geol. Surv. Prof. Paper 424–D: 248–250.

SMITH, BURNETT. 1948. Two marine Quaternary localities. Paleont. Amer. 3 (22) : 1–16.

SMITH, E. R. 1920. The Pleistocene locality at Wailes Bluff, Maryland, and its molluscan fauna. Michigan Acad. Sci. Rept. 22: 85–88.

SMITH, SANDERSON. 1867. Note on a Post-Pliocene deposit on Gardiners Island, Suffolk County, New York. Ann. New York Lyceum Nat. Hist. 8: 149–151.

STANLEY, GEORGE. 1939. Raised beaches on east coast of James and Hudson Bays (abstract). Bull. Geol. Soc. Amer. 50: 1936–1937.

STEPHENS, N. 1957. Some observations on the "interglacial" platform and the early post-glacial raised beach on the east coast of Ireland. Proc. Royal Irish Acad. 58, Sect. B, No. 6: 129–149.

STEPHENSON, L. W. 1912. Quaternary formations. In Coastal plain of North Carolina, North Carolina Geol. Surv. 3: 266–290.

STIMSON, W. 1851. Post-Pliocene deposits of Point Shirley, Massachusetts. Proc. Boston Soc. Nat. Hist. 4: 9.

SUTER, HANS. 1927. Einige Bemerkungen über die Strati-graphische Stellung der peruanischen Tablazoformation. *Centralblatt fur Mineralogie* 1927-B: 269–277.

SUTHERLAND, P. S. 1853. On the geological and glacial phenomena of the coasts of Davis Strait and Baffin's Bay. *Geol. Soc. London Quat. Jour.* 9: 296–313.

TAYLOR, G. C. 1960. Geología de la Isla de Margarita. Memoria Tercer Congresso Venezalano. 2: 838–890.

TERASMAE, JEAN. 1959a. Paleobotanical study of buried peat from the Mackenzie River Delta, Northwest Territories. *Canad. Jour. Botany* 37: 715–717.

—— 1959b. Notes on the Champlain Sea episode in the St. Lawrence lowlands, Quebec. *Science* 130: 334–335.

THORODDSEN, T. 1905. Island. Gundriss der Geographie und Geologie. *Petermanns Geog. Mitt* 32, Ergänzungsheft no. 152. Gotha.

TOLMACHOFF, I. P. 1927. Note on the discovery of the Champlain fauna on Lake St. John, Quebec, Canada. *Canadian Field-Natur.* 41: 123–125.

TOZER, E. T. 1956. Geological reconnaissance, Prince Patrick, Eglinton, and western Mellville islands, Arctic Archipelago, Northwest Territories. *Geol. Surv. Canada Paper* 55-5.

TRENCHMANN, C. T. 1925. The Scotland beds of Barbados. *Geol. Mag.* 62: 481–504.

—— 1933. The uplift of Barbados. *Geol. Mag.* 70: 19–46.

—— 1934. Tertiary and Quaternary beds of Tobago. *Geol. Mag.* 71: 481–493.

TWENHOFEL, W. S. 1952. Recent shore-line changes along the Pacific coast of Alaska. *Amer. Jour. Sci.* 250: 523–548.

UMBGROVE, J. H. F. 1939. On rhythms in the history of the earth. *Geol. Mag.* 76: 116–129.

UPSON, J. E. 1941. Late Pleistocene and Recent changes of sea level along the coast of Santa Barbara County, California. *Amer. Jour. Sci.* 247: 94–115.

—— 1951. Former marine shorelines of the Gaviota quadrangle, Santa Barbara County, California. *Jour. Geol.* 59: 415–446.

VALENTINE, JAMES W. 1961. Paleoecologic molluscan geography of the Californian Pleistocene. *Univ. Calif. Pub. Geol. Sci.* 34 (7): 309–442.

VAN DER HEIDE, S. 1957. Correlations of marine horizons in the Middle and Upper Pleistocene of the Netherlands. *Geologie en Mijnbouw* (n.w. Ser.) 19e: 272–276.

VAN DER VLERK, I. M. 1955. The significance of interglacials for the stratigraphy of the Netherlands. *Quaternaria* 2: 35–43.

VAN DER VLERK, I. M., and F. FLORSCHUTZ. 1953. The paleontological base of the subdivisions of the Pleistocene in the Netherlands. *Konink. Nederl. Akad. vam Wetenschappen afd. Natuurkunde, Vehr.* reeks 1, deel 20 (2): 1–58.

VAN VOORTHUYSEN, J. H. 1958. Foraminiferen aus dem Eemien (Riss-Würm Interglazial) in der Bohrung Amersfoort I (Locus typicus). *Meded. Geol. Stichting* 11: 27–39.

VAUGHAN, T. WAYLAND, et al. 1919. Contributions to the geology and paleontology of the Canal Zone, Panama, and geologically related areas in Central America and the West Indies. *U. S. Nat. Mus. Bull.* 103.

VAUGHAN, T. WAYLAND, C. WYTHE COOKE, W. P. WOODRING, et al. 1921. A geological reconnaissance of the Dominican Republic. *Geol. Surv. Dominican Rep. Memoir* 1.

VEATCH, OTTO, and L. W. STEPHENSON. 1911. Preliminary report on the geology of the coastal plain of Georgia. *Georgia Geol. Surv. Bull.* 26.

VEATCH, OTTO, and P. A. SMITH. 1939. Atlantic submarine valleys of the United States and the Congo submarine valley. *Geol. Soc. Amer. Special Paper* 7.

VRIES, H. DE, and A. DREIMANIS. 1960. Finite radiocarbon dates of the Port Talbot interstadial deposits in southern Ontario. *Science* 131 (3415): 1738–1739.

WAGNER, FRANCES J. E. 1959. Palaeoecology of the marine Pleistocene faunas of southwestern British Columbia. *Geol. Surv. Canada Bull.* 52.

WASHBURN, A. L. 1947. Reconnaissance geology of portions of Victoria Island and adjacent regions, Arctic Canada. *Geol. Soc. Amer. Memoir* 22.

WEAVER, C. E. 1949. Geology of the Coast Ranges immediately north of the San Francisco Bay region. *Geol. Soc. Amer. Memoir* 35.

WEISBORD, NORMAN. 1957. Notes on the geology of the Cabo Blanco area, Venezuela. *Bull. Amer. Paleont.* 38 (165): 1–25.

—— 1962. Late Cenozoic gastropods from northern Venezuela. *Bull. Amer. Paleont.* 42 (673).

WELLS, HARRY W., and HORACE G. RICHARDS. 1962. Invertebrate fauna of coquina from the Cape Hatteras region. *Jour. Paleont.* 36: 586–591.

WENNBERG, G. 1949. Differentialrörelser i inlandsisen. *Meddel. Lunds Geol. Min. Inst.* No. 114.

WENTWORTH, C. K. 1930. Sand and gravel resources of the coastal plain province of Virginia. *Virginia Geol. Surv. Bull.* 32.

WEST, R. G. 1955. The glaciations and interglacials of East Anglia and discussion of recent research. *Quaternaria* 2: 45–52. Rome.

WILSON, I. F., and V. S. ROCHA. 1955. Geology and mineral deposits of the Boleo district, Baja California, Mexico. *U. S. Geol. Surv. Prof. Paper* 273.

WILSON, J. HOWARD. 1905. The Pleistocene formation of Sankaty Head, Nantucket. *Jour. Geol.* 13: 713–734.

WOLDSTEDT, PAUL. 1955. Die Gliederung des Pleistozäns in Norddeutschland und angerenzenden Gebieten. *Geol. Fören. i Stockholm.* 77(4): 525–545.

WOLDSTEDT, PAUL, U. REIN, and W. SELLE. 1951. Untersuchungen an Nordwestdeutschen Interglazialen. *Eiszeit und Gegenwart* 1: 83–96.

WOODRING, W. P. 1924. Tectonic features of the Republic of Haiti. *Jour. Wash. Acad. Sci.* 14: 58–59.

WOODRING, W. P., M. N. BRAMLETTE, and S. W. KEW. 1946. Geology and paleontology of Palos Verdes Hills, California. *U. S. Geol. Surv. Prof. Paper* 207.

WOODWORTH, J. B., and EDWARD WIGGELSWORTH. 1934. Geography and geology of the region including Cape Cod, the Elizabeth Islands, Nantucket, Martha's Vineyard, No Man's Land and Block Island. *Mus. Comp. Zool. Memoir* 42.

WRIGHT, W. B. 1937. The Quaternary ice age. London, Macmillan.

YAKOLEV, S. A. 1926. Nanos i relef god. Leningrada i ego okrestnoceti. (Die Quatärablagerungen und Relief der Stadt Leningrad und ihrer Umgebugen.) Nauchno Melioratsionnyi Institut, *Izvestia,* Leningrad.

YANICHEVESKY, M. 1923. Kratkii predvaritel'nii otchet o geologicheskix rabotax B 41-M liste 10-verctnoi karti Evropeickoi Rosii b 1923 godu. (Compte-rendu preliminaire des recherches geologiques executée en 1923 dans la region de la feuille 41 de la carte géologique de la Russia d'Europe.) *Isv. Geol, Kom.* 1924. 43(6): 667–695.

ZAGWIJN, W. H. 1960. Aspects of the Pliocene and early Pleistocene vegetation in the Netherlands. *Meddel. Geol. Stichting.* Sec. c—III—1, No. 5: 1–78. Haarlem.

—— 1961. Vegetation, climate and radiocarbon datings in the late Pleistocene of the Netherlands. Part 1. Eemian and early Weichsellian. *Meddel. Geol. Stichting,* n.s., No. 14: 1–45. Haarlem.

ZEUNER, FREDERIC. 1948. The lower boundary of the Pleistocene. 18th Internat. Geol. Congress, London, Pt. 9: 126–130.

—— 1959. The Pleistocene period. London, Hutchinson.

ZONNEVELT, J. I. S. 1959. Litho-stratigrafische eenhenden in het Nederlandse Pleistoceen. *Meddel. Geol. Stichting,* n.s., No. 12: 31–64. Heerlen.

PLATES

PLATE 1

		PAGE
1.	*Nucula proxima* Say. Intra-Coastal Canal, Hyde County, N. C. Pleistocene. ×3	51
2–5.	*Nucula expansa* Reeve. Blow-me-down, Newfoundland. Pleistocene. ×2	51
6–8.	*Nucula tenuis* (Montagu). Yarmouth, Maine. Pleistocene. ×2	51
9.	*Nucula tenuis* (Montagu). Rivière du Loup, Que. Pleistocene. ×3	51
10–11.	*Nucula major* Richards. Cape May Canal, N. J. Pleistocene	51
12–13.	*Nuculana acuta* (Conrad). Ten miles below New Bern, N. C. Pleistocene	52
14–15.	*Nuculana pernula* (Müller). Stag Island, Que. Pleistocene	51
16.	*Nuculana minuta* (Müller). Rivière du Loup, Que. Pleistocene	52
17.	*Nuculana tenuisulcata* (Couthouy). Portland, Maine. Pleistocene	51
18.	*Yoldia glacialis* (Wood). Yarmouth, Maine. Pleistocene	52
19.	*Yoldia glacialis* (Wood). Malletts Bay, Vt. Pleistocene	52
20.	*Yoldia glacialis* (Wood). Rivière du Loup, Que. Pleistocene	52
21–22.	*Nuculana buccata* (Steenstrup). Baker Lake, N.W.T. Pleistocene. ×3	51
23–24.	*Yoldia lenticula* (Möller). Moosonee, Ont. Pleistocene. ×2	52
25.	*Yoldia limatula* (Say). Fairfield, N. C. Pleistocene	52
26.	*Yoldia myalis* Couthouy. Massachusetts. Recent	52
27–28.	*Yoldia glacialis* (Wood). Malletts Bay, Vt. Pleistocene	52
29–30.	*Nuculana jacksoni* (Gould). Portland, Maine. Pleistocene	52
31–32.	*Anadara notablils* (Röding). Snipe Key, Fla. Recent	53

(Natural size unless otherwise indicated.)

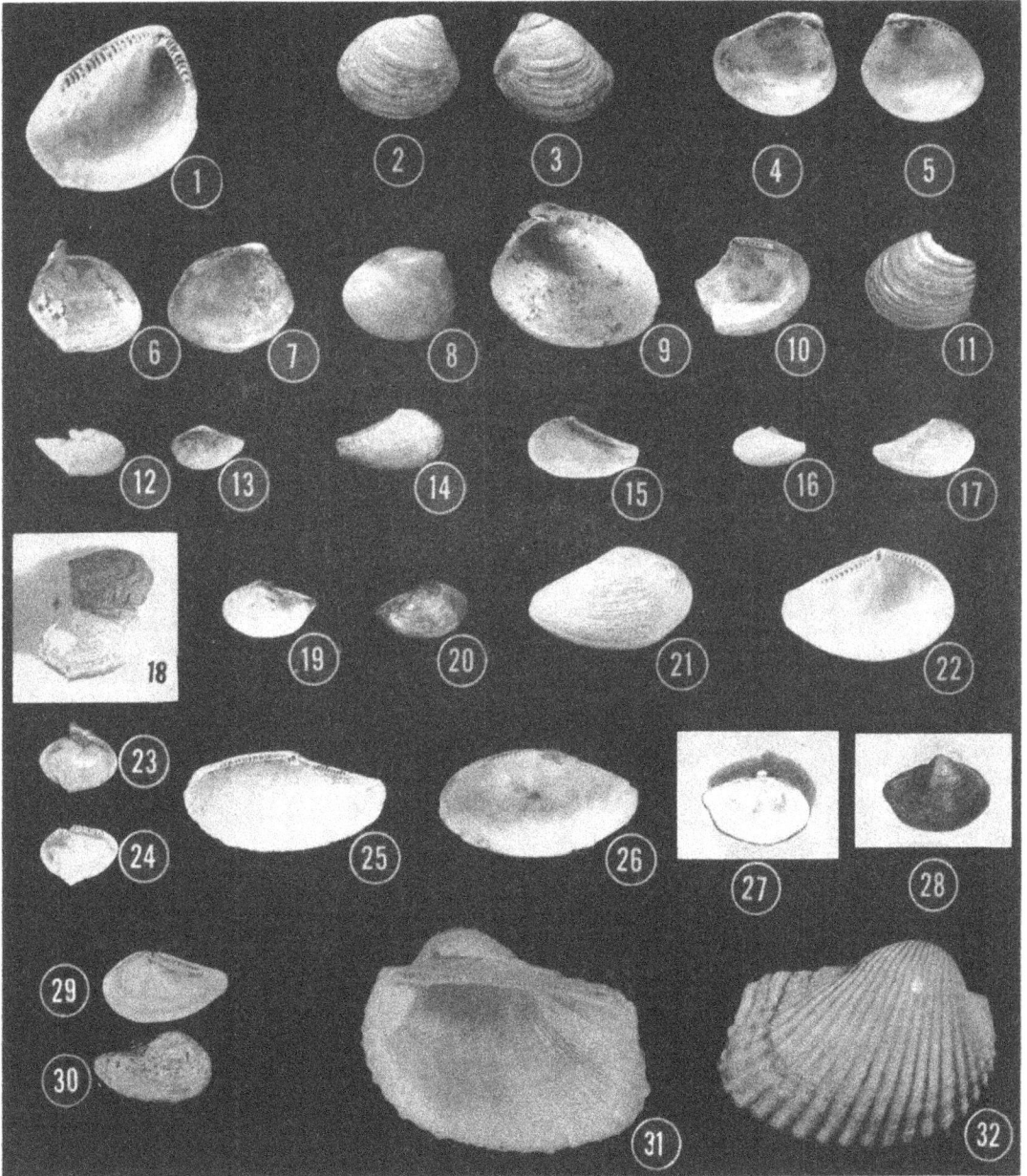

PLATE 1

97

PLATE 2

99

PLATE 3

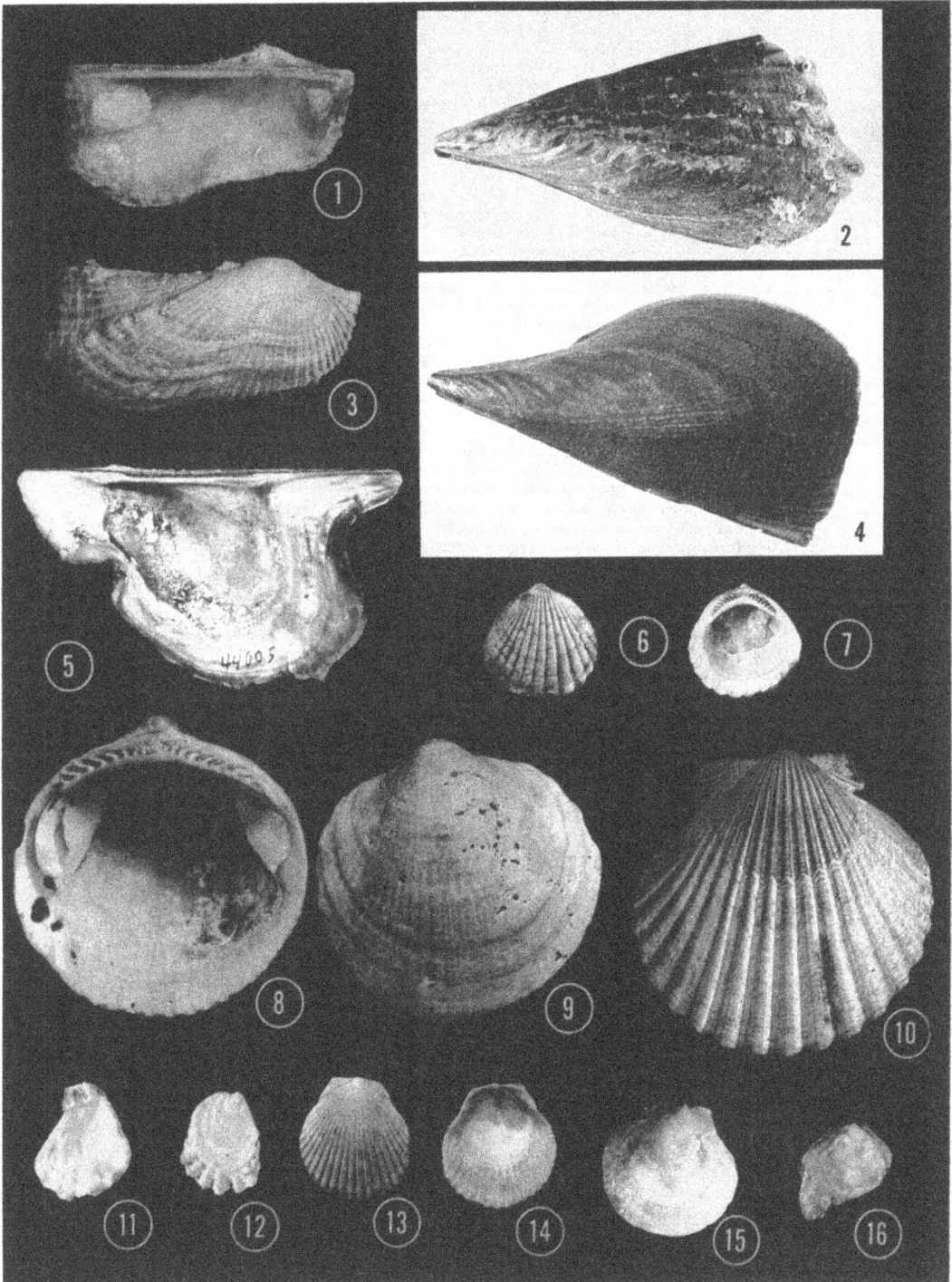

PLATE 3

101

PLATE 4

PLATE 4

103

PLATE 5

PLATE 5
105

PLATE 6

106

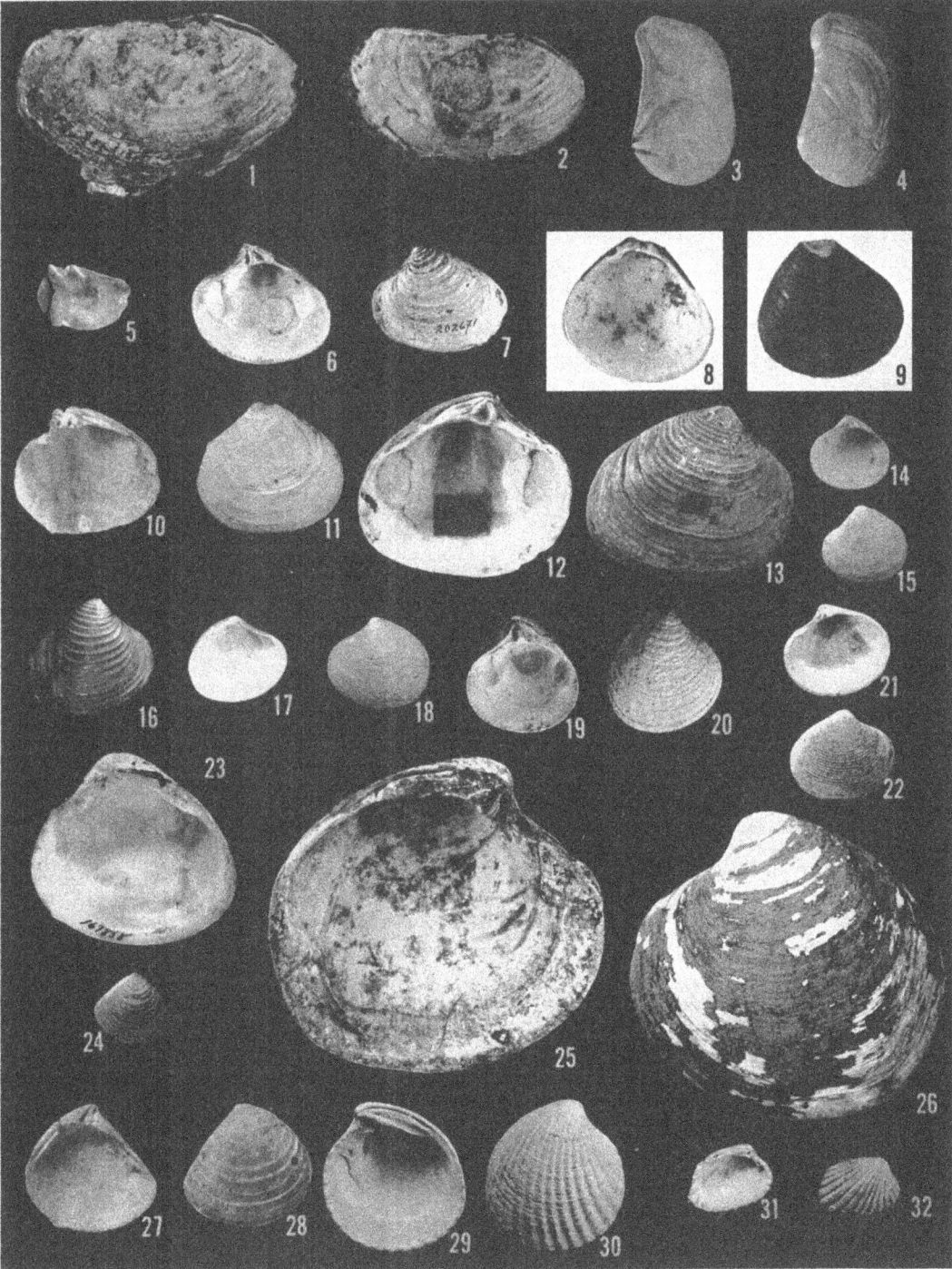

Plate 6
107

PLATE 7

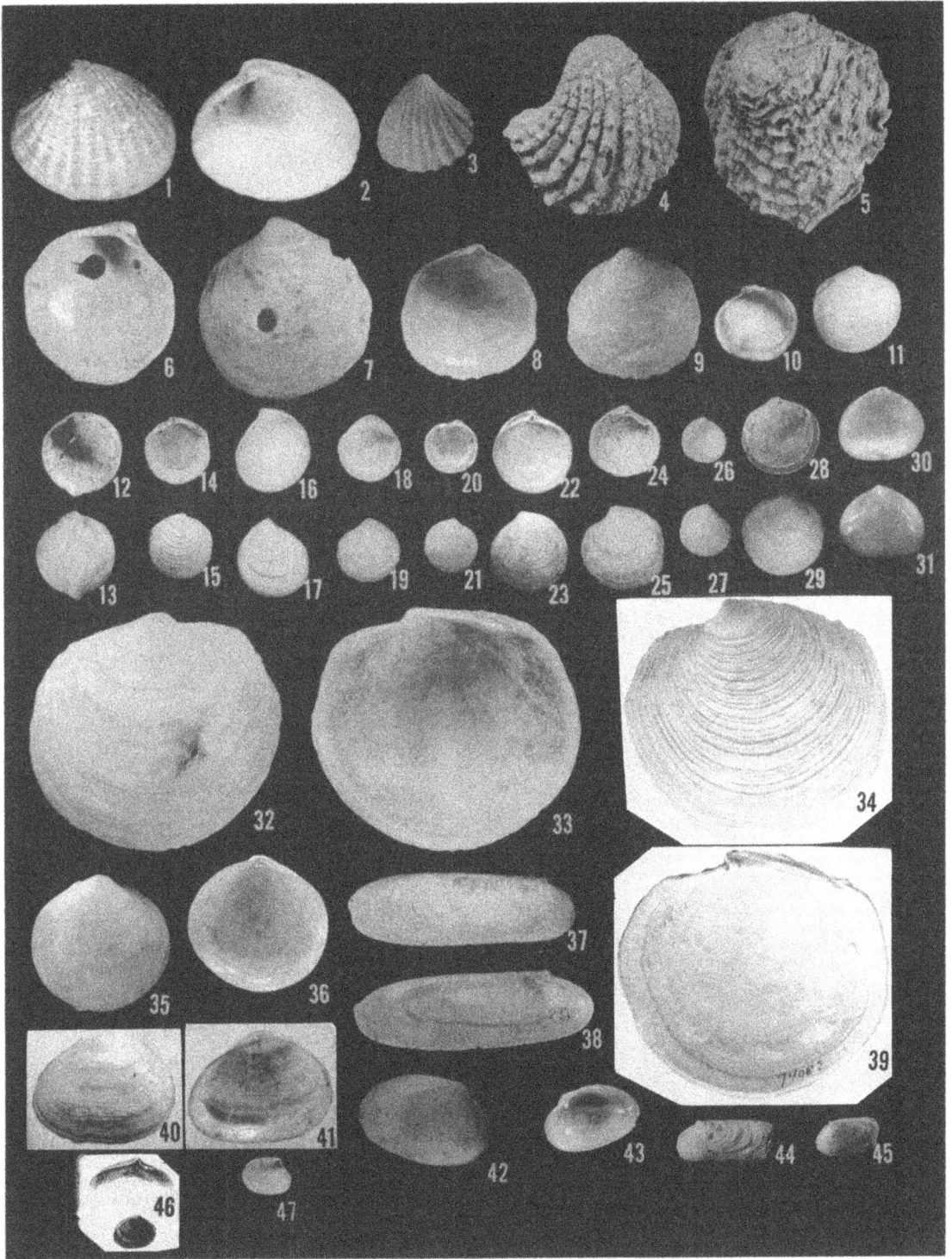

PLATE 7
109

PLATE 8

110

PLATE 8

111

PLATE 9

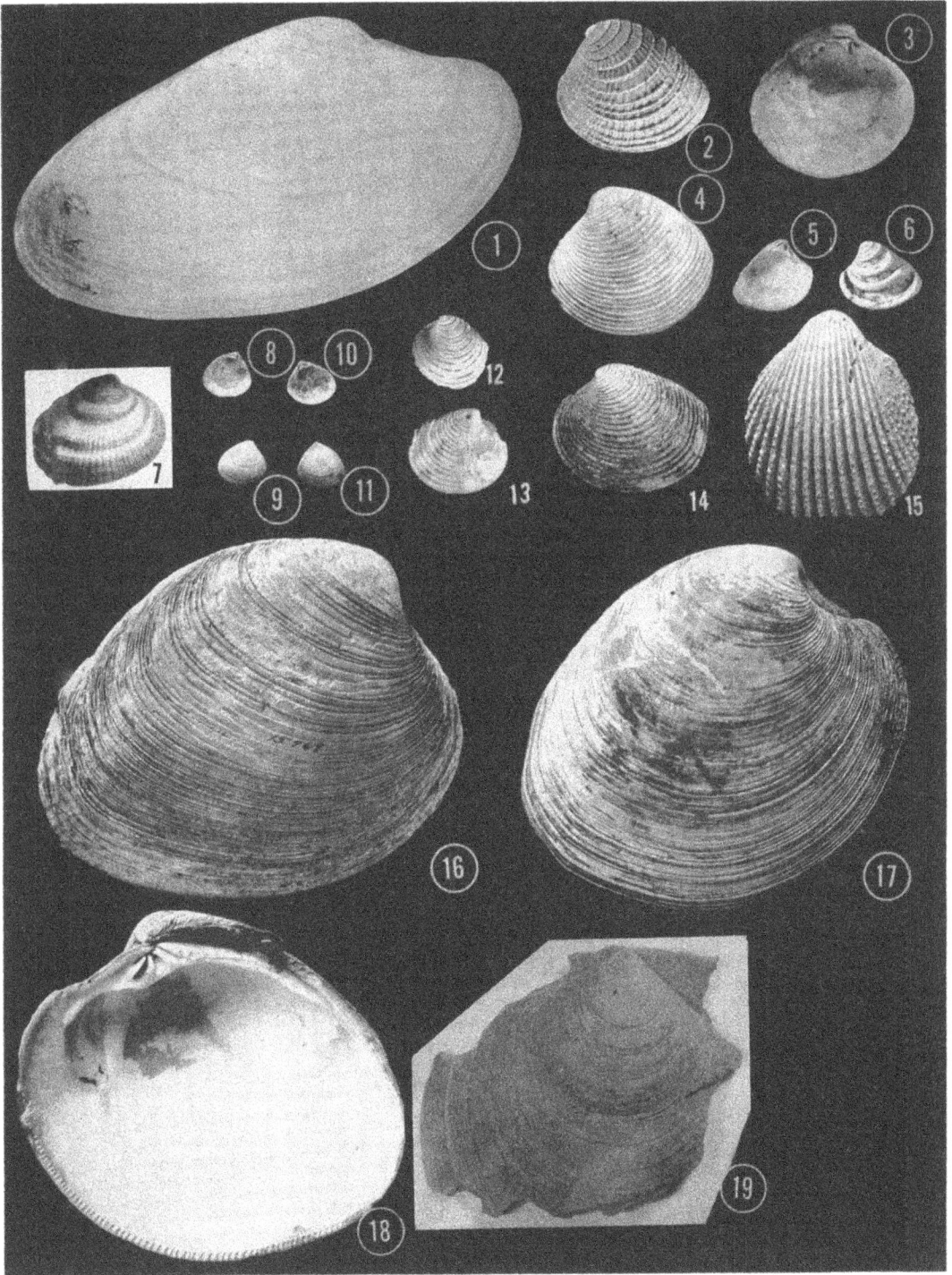

PLATE 9

113

PLATE 10

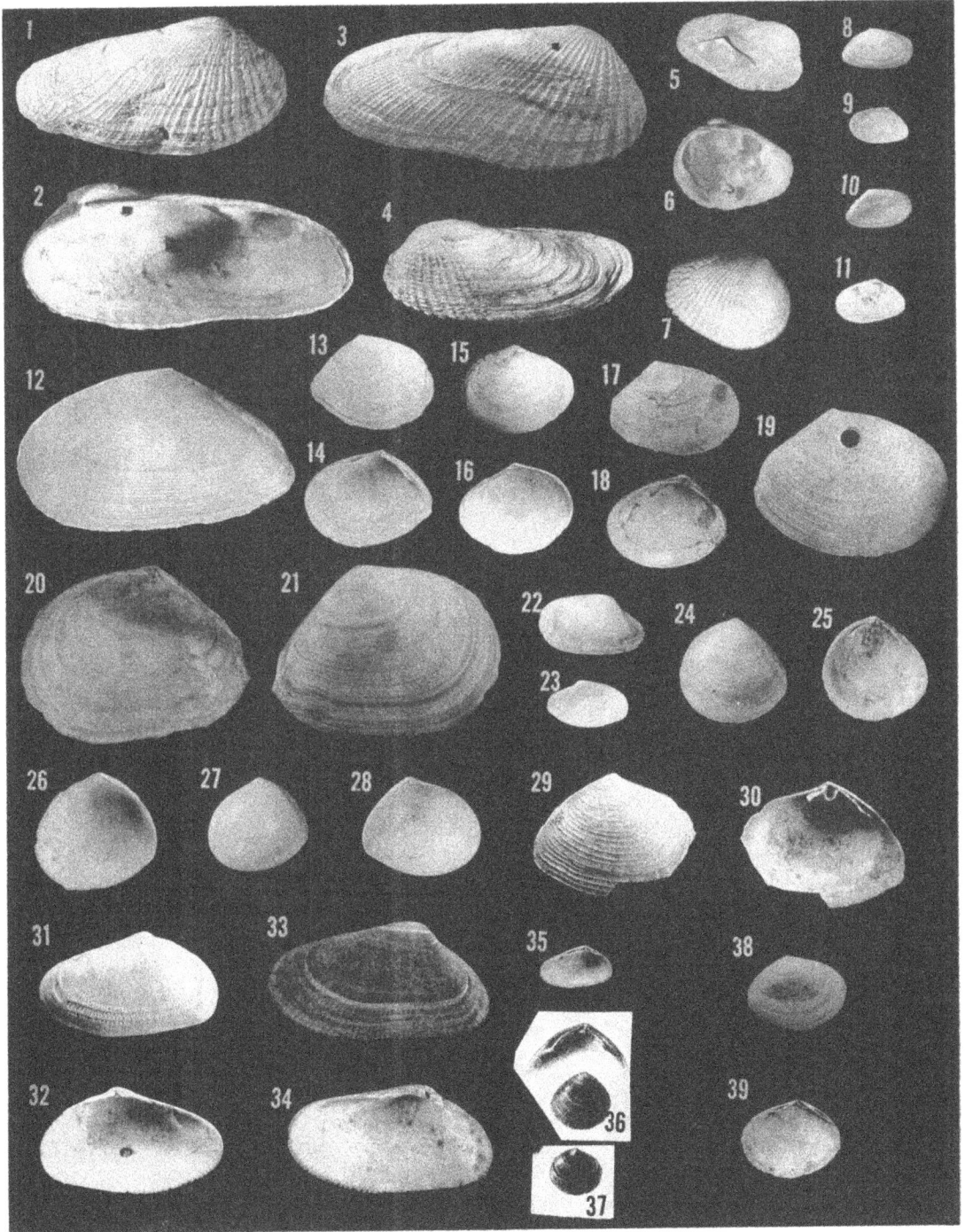

PLATE 10

115

PLATE 11

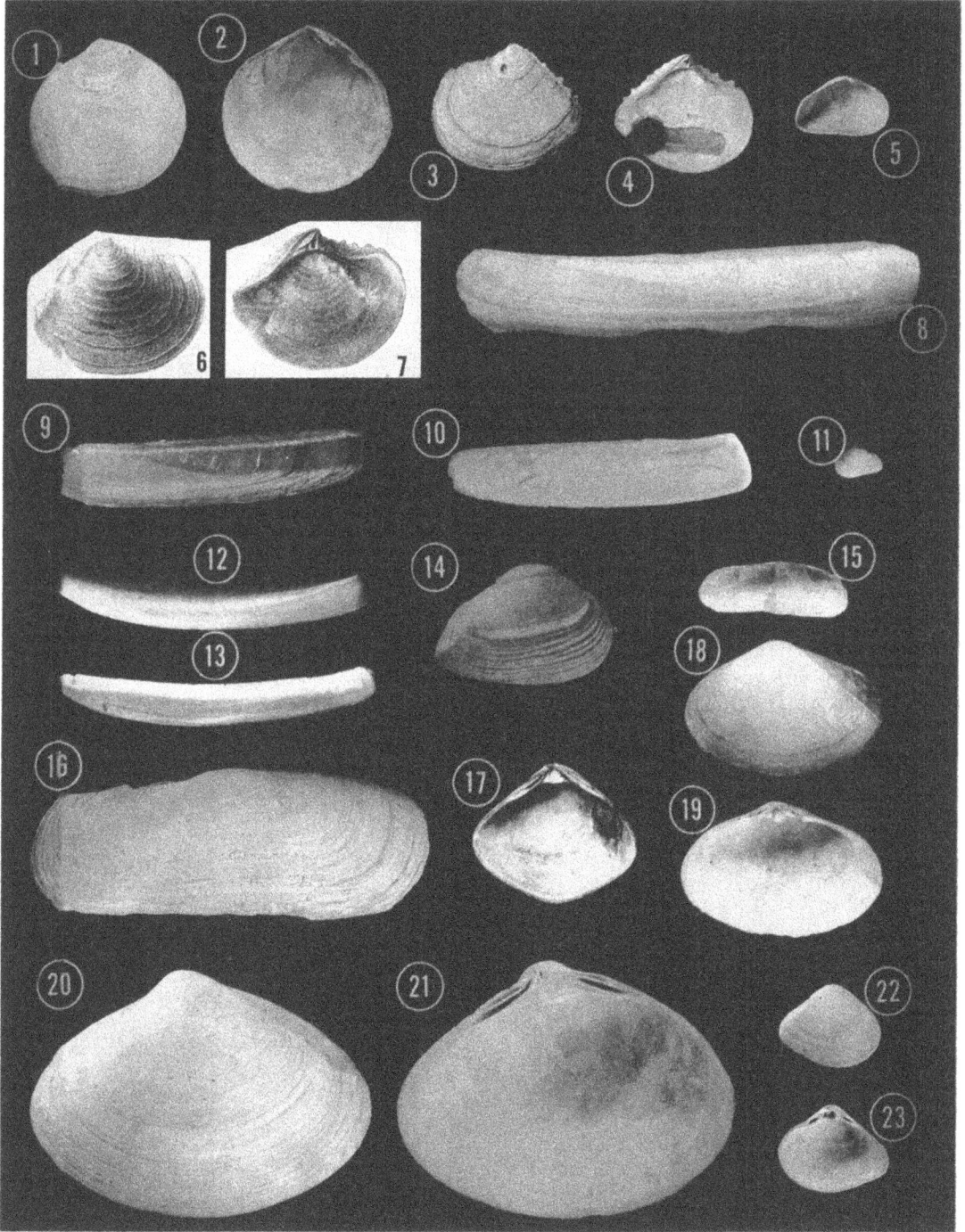

PLATE 11

117

PLATE 12

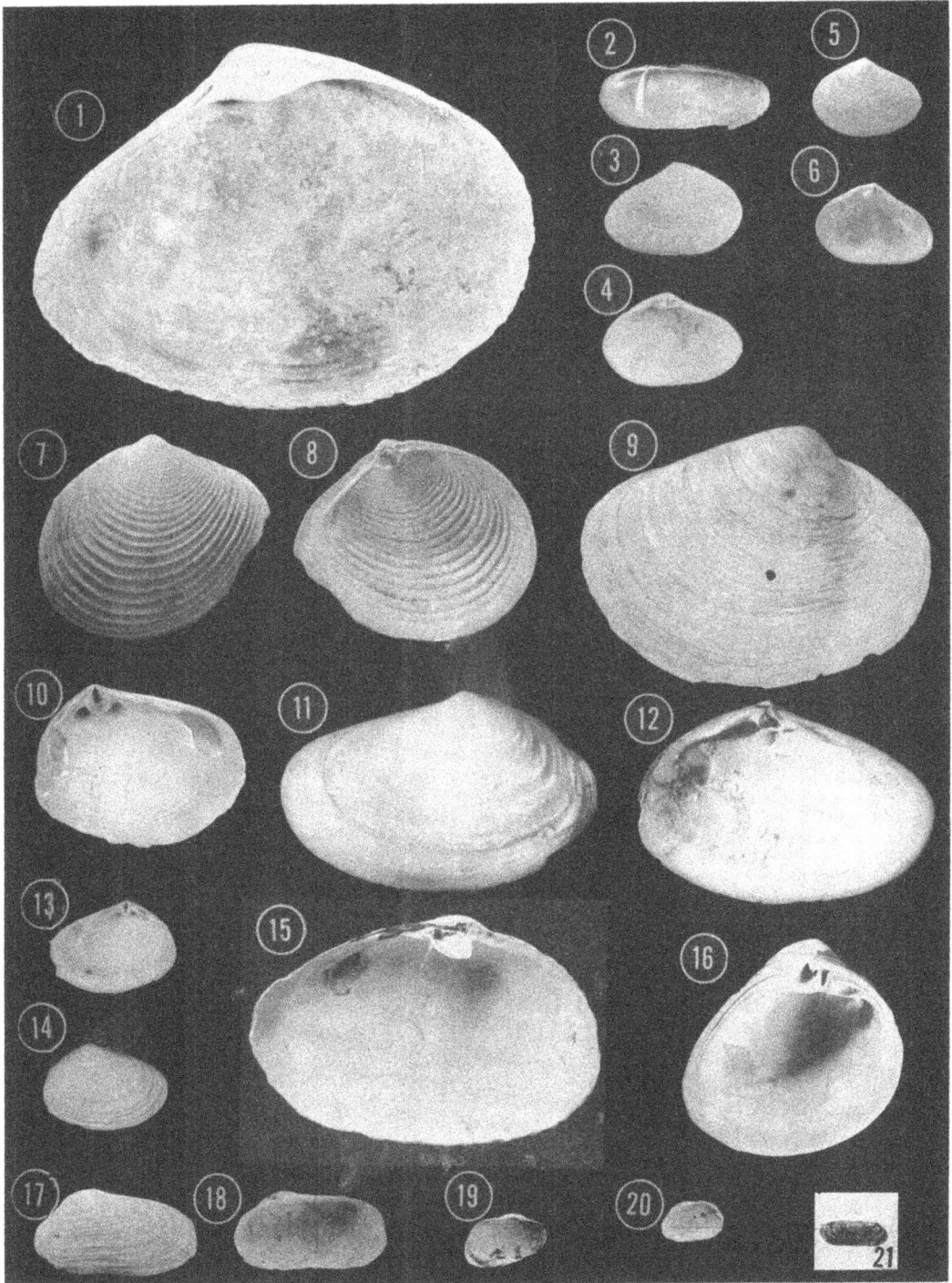

PLATE 12

119

PLATE 13

120

Plate 13
121

PLATE 14

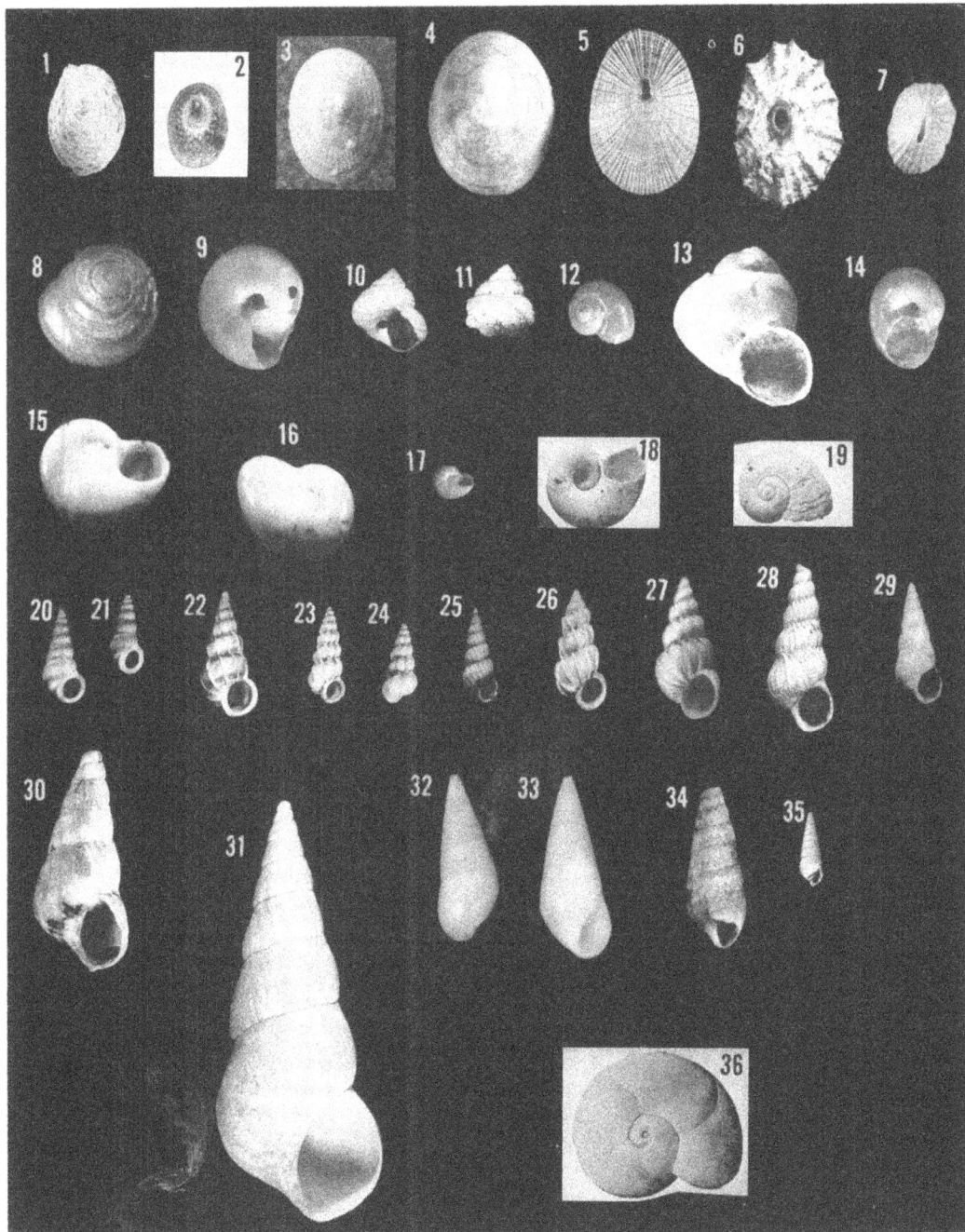

PLATE 14

123

PLATE 15

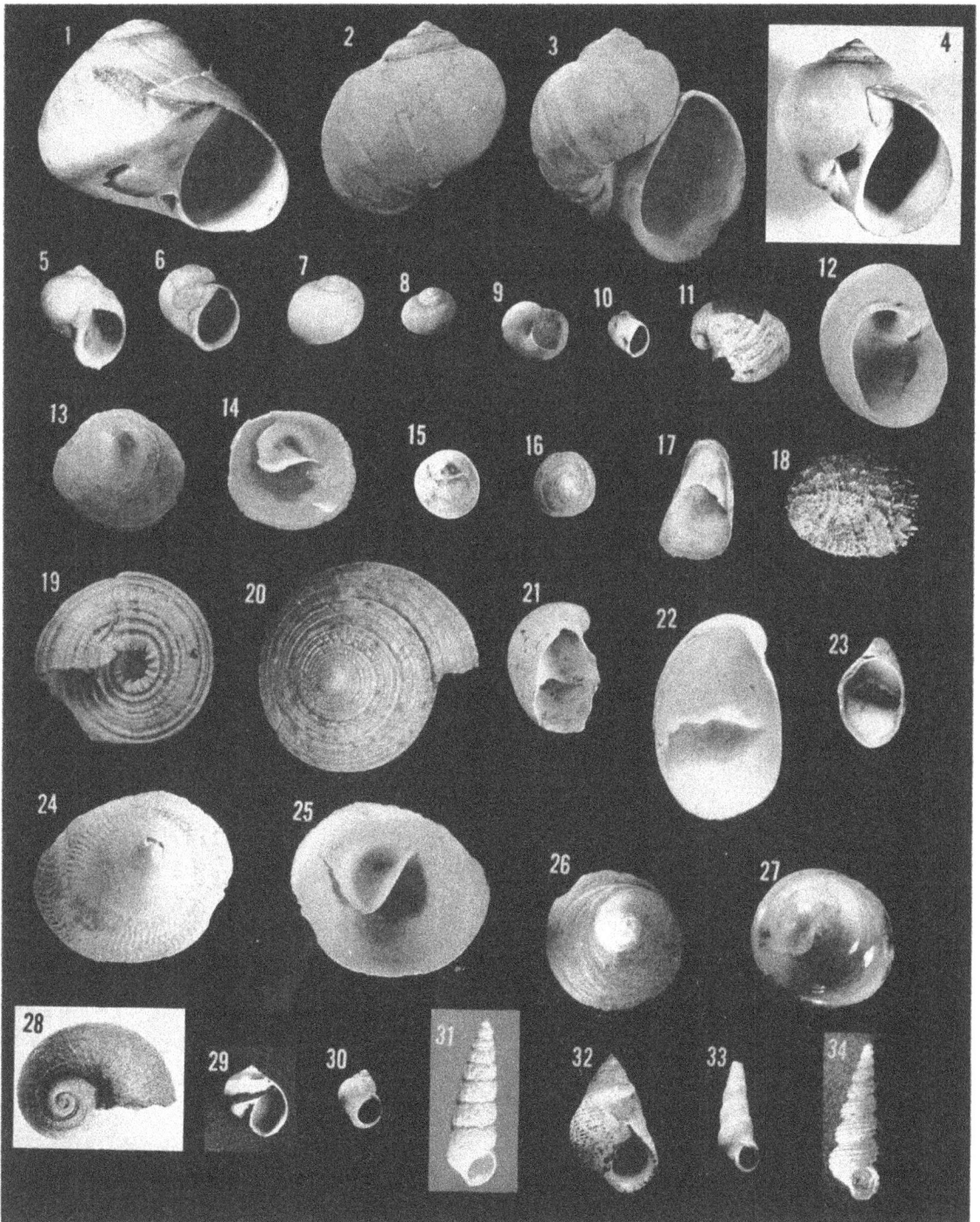

PLATE 15

125

PLATE 16

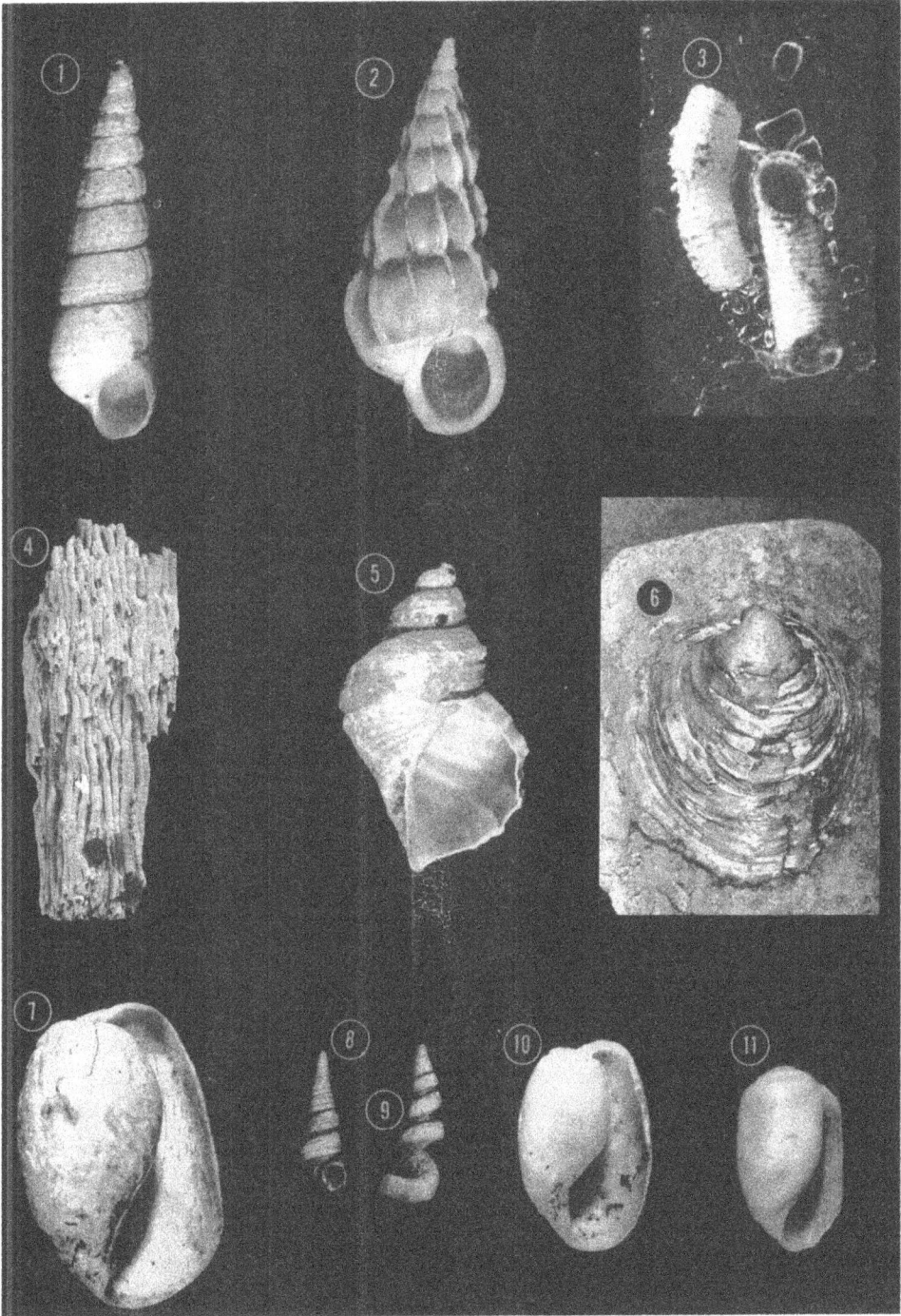

Plate 16

127

PLATE 17

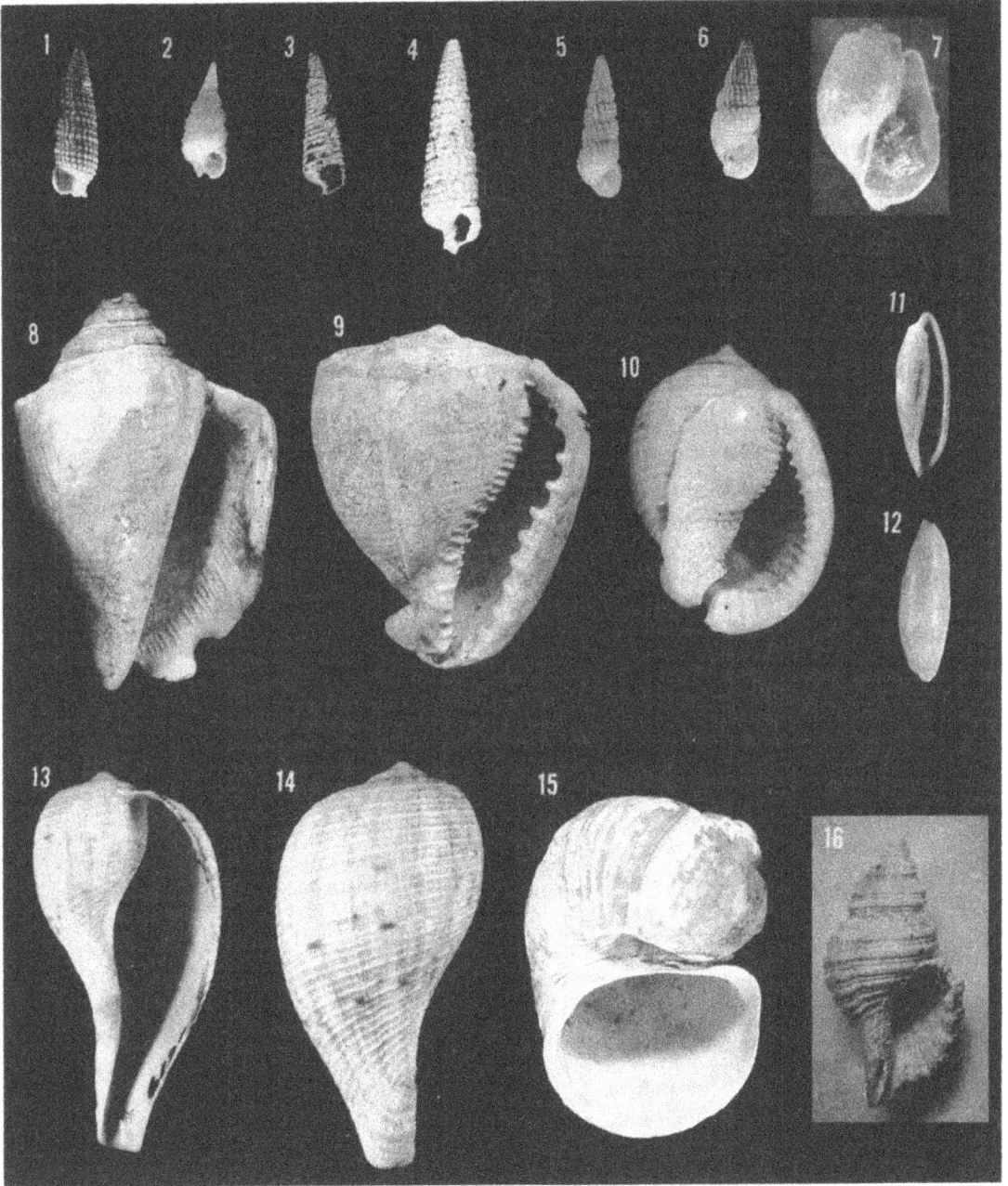

PLATE 17

129

PLATE 18

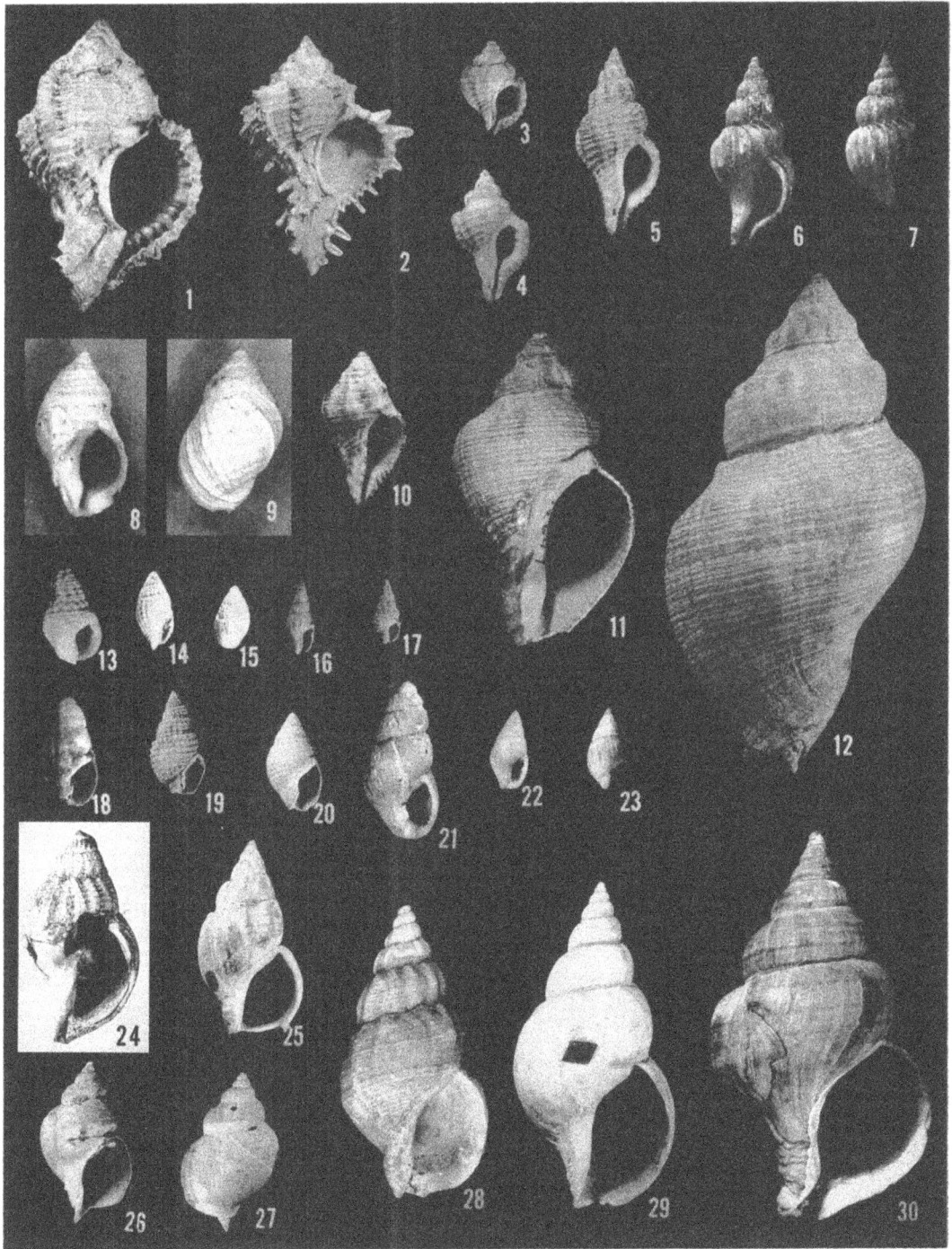

PLATE 18

131

PLATE 19

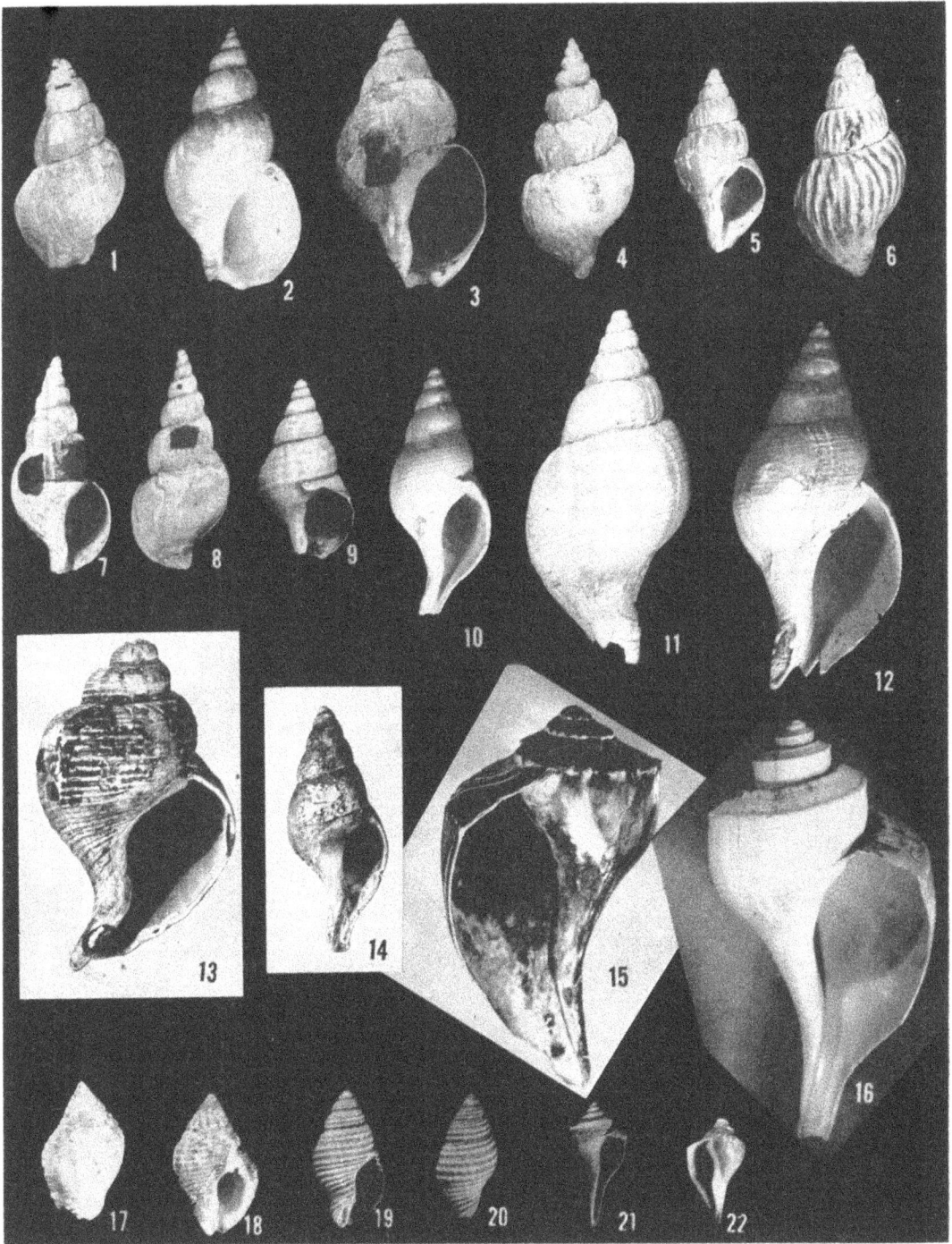

PLATE 19

133

PLATE 20

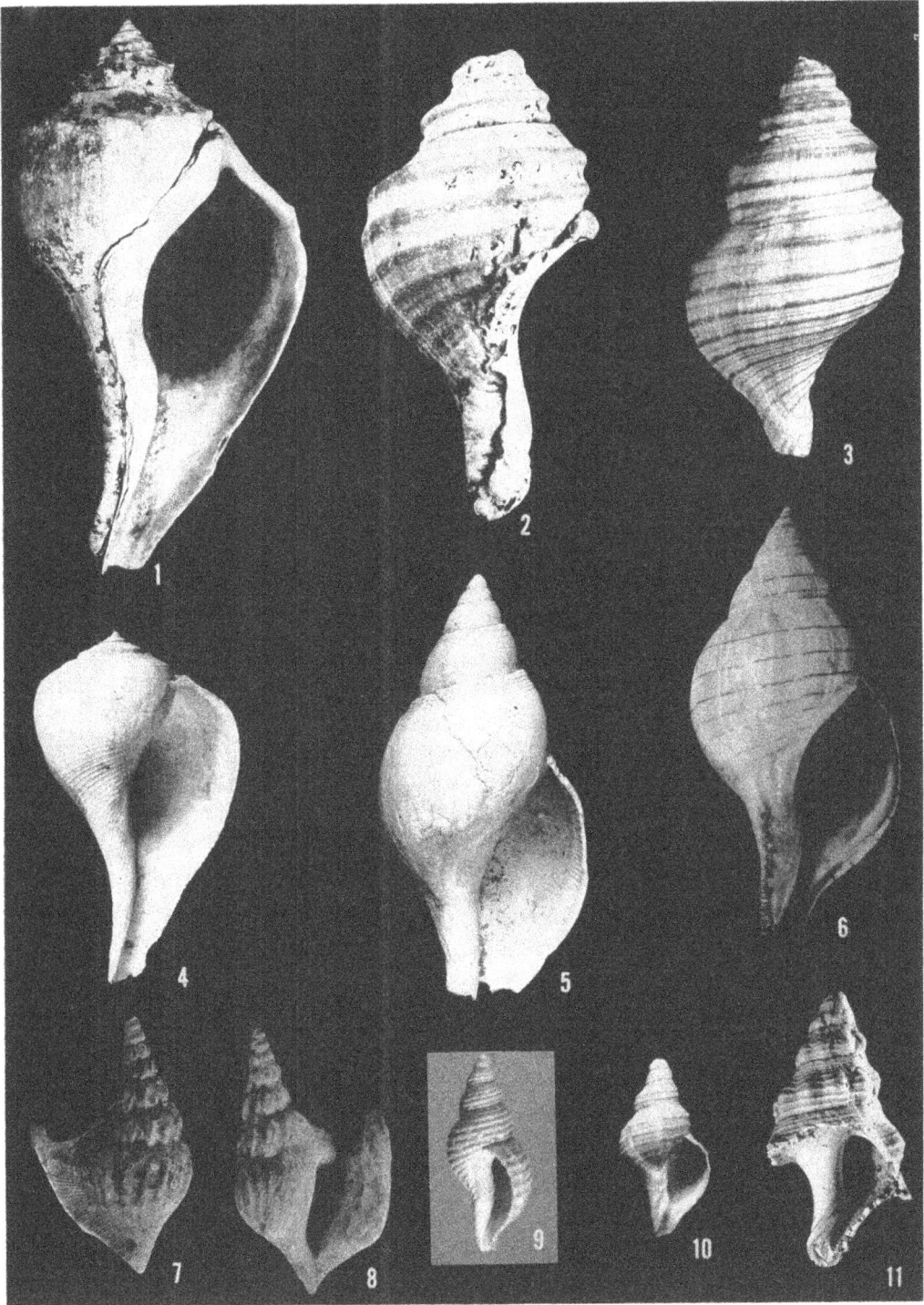

PLATE 20

135

PLATE 21

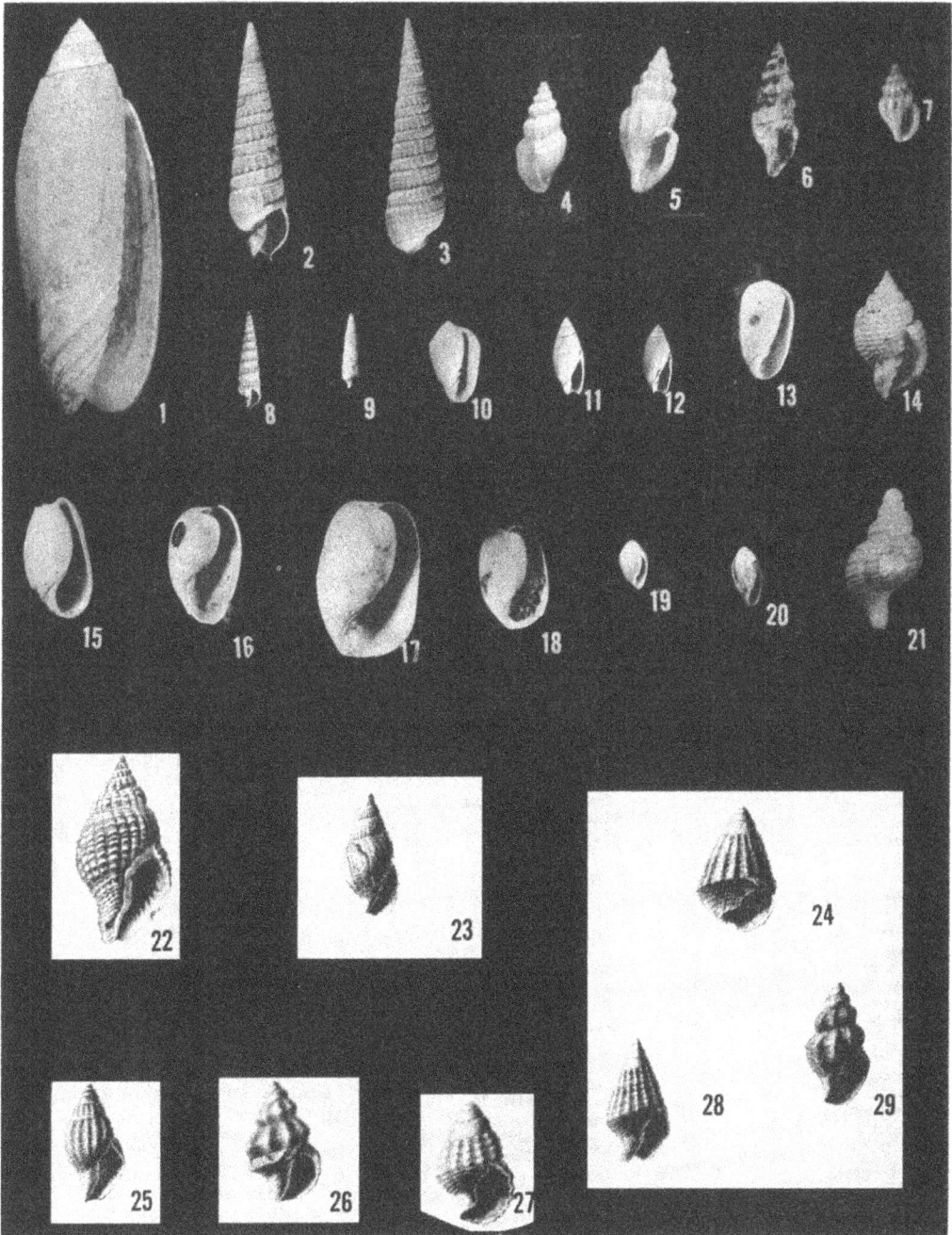

PLATE 21

137

INDEX TO PART 1

INDEX TO PART 2

(Genera and species)

conradi, Thracia, 57
contracta, Corbula, 68
contrarium, Busycon, 83
constricta, Macoma, 61
constricta, Sportella, 62
conus, Fusus, 84
convexa, Crepidula, 76
convexa, Cytherea, 64
cooperi, Caecum, 78
corrugatus, Musculus, 57
costata, Codakia, 62
costata, Siliqua, 68
costellata, Trichotropis, 78
crassa, Angaria, 77
Crassinella, 59
Crassostrea, 55
crebicostata, Astarte, 59
Crenella, 56–57
crenulata, Pyramidella, 74
Crepidula, 76–77
cribaria, Chione, 64
crispata, Zirphaea, 71
cristata, Tellidora, 67
Crucibulum, 76
cryptospira, Teinostoma, 73
Cumingia, 67
cuneata, Rangia, 69
cuneiformis, Martesia, 72
cyaneum, Buccinum, 82
Cylichna, 86
Cymatium, 79
Cyrtodaria, 70

deaurata, Mesodesma, 56
debilis, Diaphana, 86
decemcostata, Neptunea, 83
demissus, Modiolus, 56
denticulatum, Epitonium, 74
despecta, Neptunea, 83
Diaphana, 86
Dinocardium, 63
Diodora, 72
Diplodonta, 60–61
directus, Ensis, 68
discus, Dosinia, 64
dislocata, Terebra, 85
dislocatus, Pecten, 55
distans, Fasciolaria, 84
Divaricella, 62
divisus, Tagelus, 68
Donax, 67
Dosinia, 63
duplicata, Polinices, 75

Echinochama, 60
edulis, Mytilus, 56
egmontianum, Cardium, pl. 9
elegans, Choristes, 79
elegans, Dosinia, 63
elevata, Aligena, 62
elliptica, Anisodonta, 62
elliptica, Astarte, 58
Ensis, 68
Eontia, 53–54
Epitonium, 73–74
erosa, Turritella, 77
Ervilia, 70
Eupleura, 80
expansa, Nucula, 51
exustus, Brachidontes, 56

Fasciolaria, 84
Ficus, 79
filliformis, Fusus, 84
filosa, Lucina, 61
flexuosa, Strigilia, 67
floridana, Cardita, 60
floridana, Thais, 80
fornicata, Crepidula, 76
fossor, Donax, 68
fragilis, Mactra, 69
Fulgur, 83
fulvescens, Murex, 80
Fusus, 84

Gastrochaena, 71
Gemma, 65
gemma, Gemma, 65
gibbosa, Plicatula, 55
gibbus, Pecten, 55
gibbus, Tagelus, 68
gigantea, Fasciolaria, 84
glaciale, Buccinum, 82
glacialis, Yoldia, 57
glandula, Crenella, 57
Glycymeris, 54
gouldi, Bankia, 72
gouldi, Thyasira, 60
gouldiana, Pandora, 58
grandis, Pecten, 55
granulatum, Phalium, 79
granulatum, Solarium, 71
greenii, Cerithiopsis, 78
greenlandicum, Epitonium, 74
groenlandica, Macoma, 66
groenlandica, Pecten, 55
groenlandica, Polinices, 75
groenlandicus, Serripes, 63
grus, Chione, 64

Haminoea, 87
helicina, Margarites, 73
heros, Polinices, 75
Hiatella, 71
holmesii, Adeorbis, 73
humphryesii, Epitonium, 74
hyalina, Lyonsia, 57

incongrua, Arca, 53
intapurpurea, Chione, 64
intermedia, Melanella, 74
irradians, Aquipecten, 55
irrorata, Littorina, 77
islandica, Arctica, 59
islandicus, Chlamys, 55

jacksoni, Nuculana, 52

kroyeri, Buccinum, 82

Labiosa, 69
Laevicardium, 63
laevigatum, Laevicardium, 63
lamellosum, Epitonium, 73
lapillus, Thais, 80
lata, Petricola, 65
lateralis, Mulinia, 69
lateralis, Musculus, 57
latilirata, Chione, 64
laurentiana, Astarte, 59
Leda, 51–52

lens, Astarte, 59
lenticula, Yoldia, 52
Lepeta, 72
leucopheata, Congeria, 57
lienosa, Arca, 53
ligata, Fasciolaria, 84
Lima, 56
limatula, Yoldia, 52
lineata, Labiosa, 69
lineata, Scalaria, 73
lineatus, Melampus, 87
lintea, Quadrans, 60
liratulus, Colus, 83
Littorina, 77
longipes, Bornia, 62
Lora, 85
Lucina, 65
lunata, Columbella, 81
lunulata, Crassinella, 59
Lyonsia, 56

macerophylla, Chama, 60
Macoma, 66–67
Mactra, 69
Macrocallista, 64
mactracea, Crassinella, 60
madagascarensis, Cassis, 79
magellanicus, Placopecten, 55
magnum, Cardium, 63
major, Nucula, 51
Mangilia, 86
Margarites, 72–73
Marginella, 85
Martesia, 72
Melampus, 87
Melanella, 74
Mercenaria, 65
mercenaria, Mercenaria, 65
Mesodesma, 70
minor, Ensis, 68
minor, Fusus, 84
minuta, Nuculana, 52
mirabilis, Strigilia, 67
Mitra, 84
Modiolaria, 57
Modiolus, 56
modiolus, Modiolus, 56
Montacuta, 62
morruhuana, Pitar, 64
mortoni, Laevicardium, 63
multicarinata, Vitrinella, 73
multilineata, Lucina, 61
multistriatum, Epitonium, 74
Murex, 80
muricatum, Trachycardium, 63
muscosus, Aquipecten, 55
Musculus, 57
mutica, Olivella, 85
Mya, 70
myalis, Yoldia, 52
Mysella, 62
Mytilus, 56

Nassarius, 81
nassula, Phacoides, 61
Natica, 75
nautiliformis, Cochliolepis, 73
Neptunea, 83
nigra, Musculus, 57
nigricans, Vermetus, 78

www.ingramcontent.com/pod-product-compliance
Lightning Source LLC
Chambersburg PA
CBHW081337190326
41458CB00018B/6031